The Open University

Science: a level 3 course

Understanding the Continents

# Block 2

# From rifting to drifting: Mantle plumes and continental break-up

*Prepared for the Course Team by Nick Rogers*
*with contributions from Dave McGarvie*

# The S339 Course team

**Chair**

Nigel Harris

**Course Manager**

Jessica Bartlett

**Other members of the Course Team**

Mandy Anton *(Designer)*

Gerry Bearman *(Editor)*

Steve Blake *(Author)*

Steve Drury *(Author)*

Nigel Harris *(Block 4 Chair, Author)*

Martin Kemp *(BBC Producer)*

Dave McGarvie *(Author)*

Jann Matela *(Word Processing)*

Ray Munns *(Cartographer)*

Pam Owen *(Graphic Artist)*

Professor Julian Pearce, University of Wales, Cardiff *(Course Assessor)*

Nick Rogers *(Block 2 Chair, Author)*

Hazel Rymer *(Author)*

Val Russell *(original Course Manager)*

John Whalley *(Consultant Author)*

The Course Team gratefully acknowledges the contributions of those who produced the first editions of this Course, from which parts of this Block have been derived. In addition, we would like to thank the students, Associate Lecturers and assessors from other institutions who commented on drafts of this new edition. Other contributors to S339 are acknowledged in specific credits.

The Open University, Walton Hall, Milton Keynes MK7 6AA.

First published 2001. Reprinted 2002.

Edited, designed and typeset by The Open University.

Printed and bound in the United Kingdom by The Bath Press, Glasgow, UK.

ISBN 0 7492 5217 0

This block forms part of an Open University course, S339 *Understanding the Continents*. The complete list of texts which make up this course can be found at the back. Details of this and other Open University courses can be obtained from the Course Reservations and Sales Office, PO Box 724, The Open University, Milton Keynes MK7 6ZS, United Kingdom: tel. (00 44) 1908 653231. For availability of this or other course components, contact Open University Worldwide Ltd, Walton Hall, Milton Keynes MK7 6AA, United Kingdom: tel. (00 44) 1908 858585, fax (00 44) 1908 858787, e-mail ouwenq@open.ac.uk

Alternatively, much useful course information can be obtained from the Open University's website, http://www.open.ac.uk

4.1

# Contents

# 1  Introduction and study comment

Plate tectonics is the surface expression of the way the Earth loses internal heat and provides the driving mechanism for continental drift. Oceanic crust is continuously produced along the mid-ocean ridges and recycled back into the mantle at subduction zones that mark destructive plate boundaries. By contrast, continental crust, once produced, is difficult to destroy and remains on the Earth's surface where it is continually arranged and re-arranged by plate motions and mantle convection.

Today, there are seven large continental masses and a similar number of smaller, so-called microcontinents, but if the plate motions of the past 250 Ma are reversed then those various fragments can be re-assembled into one large supercontinent, Pangea. In the time since 250 Ma, Pangea has split apart, first into Gondwana and Laurasia and then into the continents with which we are familiar today. That the continents have split apart is manifest but what is the underlying cause? Continental break-up and the formation of new ocean basins is a fundamental process that shapes the surface of our planet. Yet, while plate tectonics can explain the movement of the continents and continent collision, it does not explain why large continents should break up into smaller masses. This is the problem you will be studying in this Block.

A process as complex as continental break-up cannot be fully understood using one discipline alone and that is why you will be introduced to different aspects of the geophysics, geochemistry, tectonics, petrology and geology of the areas covered. In the following four Sections, you will investigate the various stages of this process, first by looking at modern-day examples of continental break-up and then looking back through geological time to see how the principles developed from a study of these examples can be applied to a more detailed view of continental break-up in the past. The first stage of break-up involves continental rifting; the initial cracking of a continent into what may eventually become two separate continental masses. To understand this process, it is appropriate to focus on a modern-day rift where the tectonics and magmatism are still active. The best example of this is the East African Rift system, and in this Block your studies will focus on the evolution of the Kenya section of this rift. The next stage of break-up involves the transition from a continental rift into the development of a proto-ocean and the focus for this will be the geology, magmatism and tectonics of the Red Sea, the youngest ocean basin on Earth. The picture of continental break-up will then be completed with a review of the tectonic, magmatic and geodynamic processes active during the break-up of Gondwana in the Mesozoic.

However, before we tackle these challenges, you need to become more familiar with the origins and evolution of basaltic magmas. Continental break-up involves lithospheric extension and there is a close relationship between tectonic extension and the generation and eruption of basalt. Basalt is also the dominant magma generated within ocean basins. Consequently, many models of basalt generation and mantle melting have been developed from oceanic examples and the principles developed in this simpler geological environment are equally applicable to continental regimes. The following Section provides an introduction to these different models of basalt generation in the oceans and, while much of this is revision of level 2 course material, there is also new material that links basalt composition to the thermal and tectonic regimes of melt generation.

In addition to the printed text, there are numerous activities that involve calculations, hand specimen and thin section study, videocassette sequences and reading of research papers. Printed in a separate Workbook, activities are designed as an integral part of the Course and should be undertaken at the appropriate points indicated in the Block text.

# 2 The origin and evolution of basaltic magmas

## 2.1 Introduction

Partial melting of the upper mantle is the process that dominates the evolution of the outer layers of the Earth and the principal product of mantle melting is basaltic magma. Basaltic rocks, which are derived from basaltic magma, are found in both the continents and the oceans and frequently accompany tectonic activity in the lithosphere. They erupt at both constructive and destructive plate boundaries in the oceans, rifts and orogenic belts in the continents and sometimes in the centre of plates remote from any obvious tectonic activity. Significantly for this Block, basalts are often erupted in unusually large volumes during continental break-up, implying a major thermal pulse in the underlying mantle that may influence or even control continental break-up. Basalts are therefore related to both the tectonic state of the lithosphere and to the thermal structure of the underlying mantle. Quite clearly, a knowledge of the origin and evolution of basalts and basaltic magma is essential to any study of the continents.

This Section has two main purposes: first to review how basalts are generated in the oceanic environment and secondly, to provide you with a toolkit of methods and techniques that will be used to study basalts in subsequent Sections of this Block and in later Blocks. If you have successfully completed the recommended level 2 courses before starting this Course, you will find much in this Section that is familiar and may be able to work through it quite swiftly.

## 2.2 Basalts — classification

### 2.2.1 Introduction

You will have encountered basalts before, but perhaps need reminding of what a basalt is. A good working definition is a dark-coloured, igneous rock of mafic composition, with a fine-grained groundmass, and $SiO_2$ content of between 45 and 52 wt %. This definition tells us that basalt is an *igneous* rock that was once molten (totally or partly), and the dark colour ties in with the *mafic* label (rich in Fe and Mg). The *fine-grained* label tells of rapid cooling, either at the surface or in a near-surface environment (such as a dyke). This brief and simple definition provides an accurate first-order description of the most important (and abundant) volcanic rock both on Earth and on the other terrestrial (i.e. rocky) planets. However, it hides subtleties in the mineralogy and chemical composition of basalts that provide the basis for further classification and subdivisions with important implications for understanding different aspects of basalt formation. The following Sections outline three different methods of basalt classification, each of which has its own uses.

### 2.2.2 Classification by modal analysis

A newly erupted basalt consists of a combination of minerals (**phenocrysts** and/or microphenocrysts), groundmass (fine-grained or glassy) and **vesicles**. Basalts generally contain variable proportions of three important minerals: olivine, pyroxene and plagioclase feldspar. Thin section observations of rock texture (mineral shapes and sizes, whether a mineral encloses another mineral, and so on) reveal important clues about the order of crystallization by distinguishing early-formed crystals from those that formed during final solidification. Two examples of thin sections of basalts are shown in Figure 2.1. Both samples have similar bulk compositions and both have similar mineralogies, dominated by plagioclase, olivine and clinopyroxene. However, their textures are quite distinct.

- Can you recall the name of the texture in Figure 2.1a?

- Porphyritic.

The term **porphyritic** refers to samples in which a small number of large crystals are embedded in a matrix of innumerable small crystals or glass. Figure 2.1b shows a more even texture in which all crystals have roughly similar sizes and appear to interlock with each other. Such a texture is described as **intersertal**.

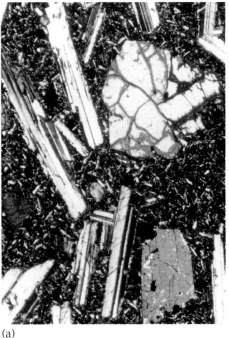

(a)                                                                        (b)

**Figure 2.1** Photomicrographs of thin sections of two basalts in cross-polarized light, showing the interference colours of the constituent minerals. These are best seen in Figure 2.1a, in which plagioclase crystals show their characteristic 'zebra-stripe' colours reflecting crystal twinning. Olivine is the virtually colourless, fractured and partly altered crystal and clinopyroxene shows red/orange/purple interference colours.

The amounts of different minerals in a rock can also be determined from a thin section — this is called a **modal analysis**. An example of a basalt modal analysis might read '… *15% plagioclase phenocrysts, 10% olivine microphenocrysts, 5% pyroxene microphenocrysts, 10% vesicles and 60% fine-grained groundmass* …'. But modal analysis has its limitations, especially if it is used to classify igneous rocks such as basalts.

- If you took a sample from the glassy rind of a pillow lava, and a sample from the more slowly cooled interior of the pillow, would you expect the modal analysis of both samples to be identical?

- No. The rapidly cooled glassy rind would contain fewer crystals than the slower-cooled interior.

This highlights a limitation in the usefulness of modal analysis for basalt classification. Simply by varying the cooling rate, it is possible to generate either a porphyritic (crystal-rich) or an **aphyric** (crystal-poor) basalt from the same starting composition. At its extreme, rapid cooling of a basalt magma will generate a glassy rock, whilst very slow cooling will generate a rock composed entirely of an interlocking network of crystals (called a gabbro).

   **Question 2.1** What do the two textures illustrated in Figure 2.1 suggest about the cooling histories of the two rocks?

The wide variation that exists in the crystal contents of basalts means that a more reliable means of classification is needed. So geologists turned to schemes based on the *chemical composition* of igneous rocks. For present purposes, we need concentrate only on two such classification schemes — using an alkali–silica plot, and using **normative minerals**.

## 2.2.3 Chemical classification

Bulk chemical analyses today form the basis of the classification of most igneous rocks. There is a myriad of different classification methods using major elements and minor elements but you only need to consider the most widely used method which is based on a plot of wt % $SiO_2$ against wt % total alkalis ($Na_2O + K_2O$). Figure 2.2 shows such a diagram and the grid used to subdivide the plot into areas that correspond to specific volcanic rock names. The main point here is not to remember the names of the different rocks but to appreciate that such volcanic rocks can be classified in a simple way by using their chemical compositions.

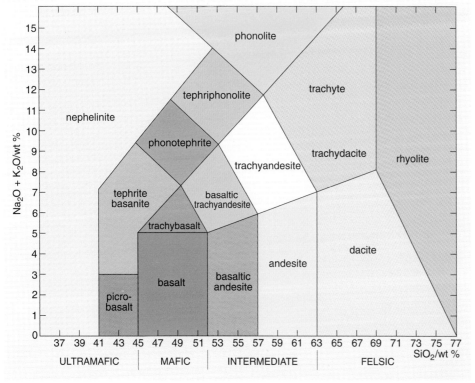

**Figure 2.2** Plot of total alkalis ($Na_2O + K_2O$) against $SiO_2$ showing a widely used subdivision of volcanic rocks.

⬤ What would you call a rock that had a total alkali content of 7% but 73 wt % $SiO_2$?

⬤ A rhyolite.

Basalt occupies a relatively small region of Figure 2.2 and is restricted to compositions with 45–52% $SiO_2$ and <5% total alkalis. Within this more restricted field, basalts are further subdivided according to their alkali and silica contents. This division is shown in Figure 2.3 and was originally based on compositional

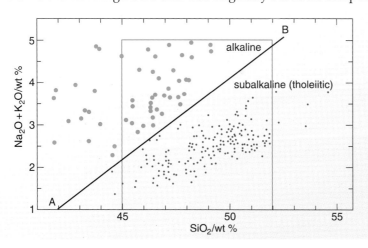

**Figure 2.3** Plot of total alkalis ($Na_2O + K_2O$) against $SiO_2$ for Hawaiian lavas. (Note that only rocks with $SiO_2$ contents between 45 and 52 wt % are basalts.) Alkali basalts plot above the A–B dividing line in the *alkali-rich* field, whilst tholeiites plot below the dividing line in the *alkali-poor* field. When this dividing line is extended to include both lower and higher $SiO_2$ compositions, it divides the alkaline and subalkaline fields. The area bounded by the red line shows the basalt field in Figure 2.2.

variations seen in basalts from Hawaii. The dividing line A–B discriminates between **alkali basalts** and **subalkaline basalts** (or **tholeiites**) (pronounced 'thoe-lee-ites'). This is an important distinction which will be developed later.

### 2.2.4 Normative classification

Put simply, a normative classification involves recalculating the major element composition of a rock into a restricted set of **normative minerals** of idealized composition. The normative minerals in basalts closely mimic the actual minerals or their solid solution end-members. Normative minerals are expressed as wt % (i.e. the same as major elements) and they are often listed beneath the major element data. You do not need to remember how to calculate norms, but Box 2.1 contains a brief description of how the calculation is done and the names and formulae of important normative minerals. A critical aspect of calculating normative minerals is the distribution of silica ($SiO_2$) between the various possible mineral compositions. An excess of silica leads to a normative mineralogy in which there is a residue of free silica after all of the other major element oxides have been assigned. By contrast, if there is a deficit of silica it is used up before all the other major oxides have been assigned. These two end results of a normative calculation lead to the description of any analysis in terms of silica saturation. An analysis with excess silica is said to be **silica-oversaturated** while one with a deficit of silica is **silica-undersaturated**. One with neither excess silica nor a deficit is **silica-saturated**.

---

### Box 2.1  Normative minerals and their calculation

Normative minerals are a set of formulae that represent ideal mineral compositions. For example, olivine can be represented as a mixture of forsterite, $Mg_2SiO_4$ and fayalite, $Fe_2SiO_4$, the pure magnesium and iron end-members of the olivine solid solution series. Similarly, the feldspars can be represented as mixtures of albite, anorthite and orthoclase. The most commonly listed normative minerals are shown in Table 2.1.

**Table 2.1**  Compositions of minerals used for normative calculations.

| Normative symbol | Mineral | Formula |
|---|---|---|
| Q | quartz | $SiO_2$ |
| C | corundum | $Al_2O_3$ |
| Or | orthoclase | $KAlSi_3O_8$ |
| Ab | albite | $NaAlSi_3O_8$ |
| An | anorthite | $Ca_2Al_2Si_2O_8$ |
| Ne | nepheline | $NaAlSiO_4$ |
| Wo | wollastonite | $CaSiO_3$ |
| En | enstatite | $MgSiO_3$ |
| Fs | ferrosilite | $FeSiO_3$ |
| Fo | forsterite | $Mg_2SiO_4$ |
| Fa | fayalite | $Fe_2SiO_4$ |
| Il | ilmenite | $FeTiO_3$ |
| Mt | magnetite | $Fe_3O_4$ |
| Ap | apatite | $Ca_3PO_4$ |

Pl (plagioclase) = Ab + An.
Di (diopside) = Wo + En + Fs
in which the ratio Wo : (En + Fs) = 1 : 1.

Hy (hypersthene) = En + Fs.
Ol (olivine) = Fo + Fa.

Clearly, the allocation of elements such as K, P and Ti is fairly straightforward because they occur only in specific minerals (orthoclase, apatite and ilmenite). By contrast, elements such as Ca, Na, Mg and Fe occur in at least two different normative minerals and so allocating the amounts of these elements has to follow a set of specific rules. The following is a brief summary of the steps required to recast a major element analysis into its normative constituents. You are not required to perform these calculations and this summary is for background information only.

*Step 1*  Major element oxide data in wt % are converted to molecular proportions by dividing each value by its relative molecular mass.

*Step 2*  Normative accessory minerals (such as apatite, ilmenite and magnetite) are calculated first by allocating all the $P_2O_5$, $TiO_2$ and $Fe_2O_3$ and the necessary CaO and FeO to these minerals.

*Step 3*  The distribution of the alkali elements, $Na_2O$ and $K_2O$, and aluminium, $Al_2O_3$, is established by allocating them and the appropriate amount of $SiO_2$ to the feldspars (orthoclase and albite). Any remaining $Al_2O_3$ is combined with CaO and $SiO_2$ to calculate normative anorthite.

*Step 4*  Any remaining CaO, MgO and FeO are allocated to the pyroxene normative minerals (wollastonite, enstatite and ferrosilite), together with the necessary $SiO_2$. These components are then combined to determine normative clinopyroxene or diopside (wollastonite, enstatite and ferrosilite in the proportion Wo: (En + Fs) = 1 : 1) and orthopyroxene or hypersthene (enstatite + ferrosilite) from any excess in En and Fs.

*Step 5*  At this point, there is either an excess of silica or a deficit. If there is an excess, it is listed as normative quartz, Q. If there is a silica deficit, then as much orthopyroxene is converted to olivine as is necessary to balance the excess magnesium and iron. In some cases, all of the orthopyroxene will be converted to olivine and the normative mineralogy will be free of hypersthene. If after this step there is still a deficit of silica, yet more silica is taken from albite which is converted to nepheline until the silica deficiency is finally balanced. **Nepheline** is an alumino-silicate with a framework structure, like a feldspar but with less silica — hence it is called a **feldspathoid**. In thin section, it is clear, colourless and has low relief and can be easily confused with quartz. Like albite, the main cation is sodium and there is a reaction relation between silica and nepheline to produce albite:

$$2SiO_2 + NaAlSiO_4 = NaAlSi_3O_8.$$
$$\text{quartz} \quad \text{nepheline} \qquad \text{albite}$$

Hence, quartz and nepheline are never seen to co-exist in rocks.

It is step 5 that is most critical in classifying basaltic compositions as these have low silica contents but high CaO, FeO and MgO. If you follow through step 5 closely, you will appreciate that certain minerals can appear together while others are mutually exclusive. Quartz can co-exist with hypersthene and hypersthene can co-exist with olivine but olivine cannot co-exist with quartz. Similarly, nepheline can co-exist with olivine but not with hypersthene or quartz. These discrete normative associations mirror those seen in the mineralogy of real basalts and associated rocks, and are used as the basis of the normative classification.

Now look at Table 2.2. This shows how a basalt is classified by its normative mineralogy.

Table 2.2  Summary of important relationships between silica saturation, diagnostic normative minerals, and basalt rock name. The diagnostic normative minerals are particularly important. (The other main normative minerals are listed, but these are of lesser importance.) The 'saturated' basalt does not have its own diagnostic minerals — rather the *absence* of both quartz and nepheline is diagnostic.

| Rock name | Silica saturation | Diagnostic normative mineral(s) | Other main normative minerals | Field occupied on $SiO_2$ vs. $Na_2O + K_2O$ plot (Figure 2.3) |
|---|---|---|---|---|
| quartz tholeiite | oversaturated | quartz (Q) | hypersthene (Hy) | *subalkaline* |
| olivine tholeiite | saturated | absence of both quartz (Q) and nepheline (Ne) | olivine (Ol), pyroxene (Di + Hy) | *subalkaline* |
| alkali basalt | undersaturated | nepheline (Ne) | olivine (Ol), pyroxene (Di) | *alkaline* |

● What is the major difference between the basalt names on Figure 2.3 (classified using the alkali–silica plot) and the basalt names on Table 2.2 (classified using normative minerals)?

● On Figure 2.3, there is a two-fold division, into alkali basalts and tholeiites. On Table 2.2, there is a three-fold classification, into alkali basalts, olivine tholeiites, and quartz tholeiites.

You have already encountered the names 'alkali basalt' and 'tholeiite' when classifying basalt using the alkali–silica plot (Figure 2.3), but why are there now two 'tholeiites'? One of the central planks of determining the normative mineralogy of a rock is its degree of silica saturation. During the normative mineral calculation, one of the final steps involves the distribution of $SiO_2$ between the minerals olivine, hypersthene, albite, nepheline and quartz. This is done in such a way that a normative analysis cannot contain quartz together with either olivine or nepheline. There is a sound chemical basis for this that you will read more about in the next Section. The consequence is that, as with real basalts, a normative analysis of a basalt can contain Q + Hy, Ol + Hy, or Ol + Ne. In addition to these components, each normative analysis also includes Ab, An, and Di. These three distinct normative mineralogies lead to three different types of basalt, known as **quartz tholeiite** (Q + Hy), **olivine tholeiite** (Hy + Ol) and **alkali basalt** (Ol + Ne). In addition, quartz tholeiites are also said to be **silica-oversaturated**, olivine tholeiites are **silica-saturated** and alkali basalts are **silica-undersaturated**.

● Which normative mineral is common to all tholeiites?

● Hypersthene (Hy).

To help illustrate these concepts further, you should now attempt Question 2.2 which involves the major element and normative analyses of the three basalts listed in Table 2.3.

> **Question 2.2** Classify the three basalts in Table 2.3 using (a) their bulk compositions and (b) normative mineralogy. (c) Comment on the compatibility of these two classification methods.

Table 2.3   Bulk composition and normative mineralogy of three basalts.

| Major element oxides | Basalt 1 | Basalt 2 | Basalt 3 |
|---|---|---|---|
| $SiO_2$ | 47.23 | 48.88 | 51.08 |
| $TiO_2$ | 2.41 | 1.09 | 2.35 |
| $Al_2O_3$ | 15.17 | 16.59 | 13.81 |
| $Fe_2O_3$ | 1.50 | 1.50 | 1.80 |
| FeO | 10.40 | 7.35 | 11.91 |
| MnO | 0.15 | 0.24 | 0.34 |
| MgO | 6.03 | 7.37 | 5.76 |
| CaO | 9.54 | 12.66 | 7.64 |
| $Na_2O$ | 2.91 | 2.42 | 2.47 |
| $K_2O$ | 1.91 | 0.19 | 0.99 |
| $H_2O$ | 2.41 | 1.56 | 1.52 |
| $P_2O_5$ | 0.34 | 0.15 | 0.29 |
| total | 100.00 | 100.00 | 99.96 |
| norms | | | |
| Q | – | – | 3.64 |
| Or | 11.2 3 | 1.12 | 5.87 |
| Ab | 20.72 | 20.48 | 20.89 |
| An | 22.69 | 33.85 | 23.67 |
| Ne | 2.12 | – | – |
| Di | 18.51 | 22.77 | 10.31 |
| Hy | – | 8.86 | 26.33 |
| Ol | 14.73 | 6.78 | – |
| Mt | 2.18 | 2.18 | 2.61 |
| Il | 4.58 | 2.07 | 4.46 |
| Ap | 0.80 | 0.35 | 0.68 |

In summary, using their chemical composition we can conveniently classify basalts into two broad types: alkali basalts (which have high $Na_2O + K_2O$ and are Ne-normative) and tholeiites (which have low $Na_2O + K_2O$ and are Hy-normative). Tholeiites can be further subdivided into olivine tholeiites and quartz tholeiites using their normative mineralogy. In the next Section, we will explore further the importance of the distinction between alkali basalts and tholeiites, and how this has profound consequences for the crystallization of liquids derived from these two different types of basalt magma.

---

### Activity 2.1    Basalts in thin section

Now look at *Workbook 2* and follow the instructions for this Activity, which takes you through some aspects of studying basaltic rocks in thin section. You should take about 90 minutes to do this Activity.

---

## 2.2.5   Summary of Section 2.2

- A basalt is defined as a dark, fine-grained mafic igneous rock with a $SiO_2$ content between 45 and 52 wt % and less than 5% $Na_2O + K_2O$.

- Modal analysis of erupted rocks such as basalts has limited use due to wide variations in crystal content resulting from local cooling conditions rather than being a fundamental property of the magma.

- An alkali–silica plot ($Na_2O + K_2O$ versus $SiO_2$) effectively distinguishes alkali basalts from subalkaline basalts.

- Normative analysis takes the major element composition of a basalt and re-casts this into a set of standard anhydrous minerals.

- Normative analysis classifies basalts by their degree of silica saturation: alkali basalts (undersaturated and Ne-normative), olivine tholeiites (saturated and Hy-normative), and quartz tholeiites (oversaturated and Q + Hy-normative).

- Basalts which plot as alkali basalts on an alkali–silica plot will usually be classified as alkali basalts using their normative mineralogy. The same applies to tholeiitic basalts, although normative classification allows their further subdivision into quartz and olivine tholeiites.

# 2.3   Basalts — petrogenesis

## 2.3.1   Basalts and oceanic environments

The subdivision of basalts into different categories depending upon their degree of silica saturation is of fundamental importance because alkali basalts and tholeiites tend to be characteristic of different tectonic environments. Basalt is the dominant magma type in the oceans and this is the best environment in which to start our investigation.

- Can you suggest three oceanic locations where basaltic volcanism occurs?

- Mid-ocean ridges (i.e. constructive plate boundaries); ocean islands (i.e. within-plate volcanism); and island arcs (destructive plate boundaries).

You will read more about the melting processes which produce the basalts that are erupted in these different locations later, but for now it is sufficient to remember these three different locations.

How do basalt compositions vary between these different locations? Figure 2.4 shows compositional data from three examples.

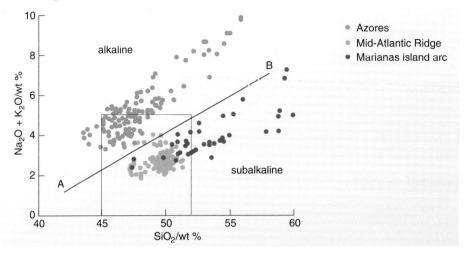

**Figure 2.4** Alkali–silica plot of basalts (45–52% $SiO_2$), and associated rocks, from various oceanic locations. The Azores from the central Atlantic represent an ocean island setting (within-plate), the Mid-Atlantic Ridge represents a constructive plate boundary, and the Marianas island arc in the western Pacific represents a destructive plate boundary. The line A–B divides alkali basalts (above) from subalkaline basalts (below).

**Question 2.3**  Given the information in Figures 2.3 and 2.4, can you draw any conclusions regarding the relationship between basalt type and tectonic regime?

○  What other important difference is evident between the rocks of the Azores and the Mid-Atlantic Ridge on Figure 2.4?

○  All of the Mid-Atlantic Ridge rocks are basalts ($SiO_2$ 45–52%), whereas a wider range of rock compositions is present in the Azores (from $c.$ 43–56% $SiO_2$).

**Question 2.4**  (a) What would you call the two ocean island rocks on Figure 2.4 that have the highest $SiO_2$ and the highest total alkali content? (b) What would you call the island arc rock on Figure 2.4 with the highest $SiO_2$ content?

So, not only is there a broad distinction between the basalts erupted at ocean ridges and ocean islands, but there also appears to be a wider range of compositions erupted in ocean islands compared to mid-ocean ridges (as measured by the spread in $SiO_2$ content). Furthermore, Figure 2.4 reveals that the higher-silica rocks associated with a basalt type lie on the same side of the dividing line as the basalt. This observation suggests that the dividing line between alkaline and subalkaline basalts exerts some influence on the pathways followed by high-silica magmas. The reasons for this distinction lie in the crystallization history of basaltic magmas that can be explored using phase diagrams.

## 2.3.2  Phase diagrams — introduction

Phase diagrams have been introduced in earlier Earth science courses although you may need reminding of some of the principles involved in their interpretation. Activity 2.2 is a revision exercise that explains how to interpret both binary and ternary phase diagrams and revises the concepts of solid solution and eutectic compositions.

## Activity 2.2  Revision of phase diagrams

This Activity takes you through a number of exercises to explain the principles involved in the interpretation of phase diagrams, and should take you about one hour to complete.

For a phase diagram to be useful in understanding the evolution of basalts, it must be an approximate analogy for a natural basalt magma.

● What are the three main minerals that most commonly crystallize from a basalt magma?

● Olivine, pyroxene and plagioclase feldspar.

So we need a phase diagram that contains olivine, pyroxene, and plagioclase feldspar (or their normative analogues). But, as you have seen, the two-fold division into alkali basalts and tholeiites also requires the addition of other normative minerals.

● What is the characteristic normative mineral of first, a quartz tholeiite and secondly, an alkali basalt?

● Quartz (Q) and nepheline (Ne).

To be an analogy for natural basalt magma, the 'ideal' phase diagram needs to represent five normative components: olivine (Fo), pyroxene (Di), plagioclase feldspar (Ab), quartz (Q), and nepheline (Ne). Rather than jump straight to this, we will start by considering a binary phase diagram (Ne–Q), and then consider a ternary phase diagram (Ne–Q–Di).

### 2.3.3  Binary phase diagram (Ne–Q)

Figure 2.5 is a phase diagram which, although a binary (two-component Ne–Q) diagram, illustrates relationships between *three* components: nepheline, quartz and albite. Albite (a plagioclase feldspar) is formed when nepheline and quartz react with each other according to the reaction:

$$NaAlSiO_4 \text{ (nepheline)} + 2SiO_2 \text{ (silica)} = NaAlSi_3O_8 \text{ (albite)}.$$

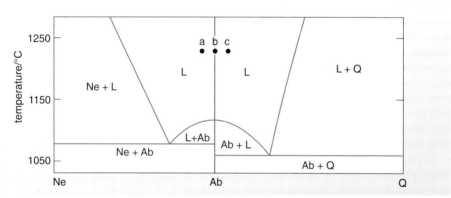

**Figure 2.5**  The binary nepheline (Ne) – silica (Q) system at low pressure (0.1 MPa = atmospheric pressure), with Ab representing the intermediate compound albite, and L representing liquid.

Although Figure 2.5 looks complex, it can be divided into two halves by a vertical line (shown in red) at the composition of Ab (albite). To the right of this line, there is a binary eutectic between albite and quartz, and to the left a binary eutectic between nepheline and albite. Given this information, you should now attempt Question 2.5.

**Question 2.5**  (a) What will be the sequence and final product of crystallization of melts with the compositions labelled a and c in Figure 2.5? (b) Under what conditions (if any) could a melt with a composition between albite and quartz (Q) give rise to a liquid containing nepheline? (c) What will be the final product of crystallization of a liquid with the composition b in Figure 2.5?

### 2.3.4  Ternary phase diagram (Ne–Q–Di)

Now we can move from a binary to a ternary phase diagram, Ne–Q–Di (Figure 2.6), by bringing a fourth component — diopside (pyroxene) — into consideration. Once again, the three different compositions (a, b and c) are shown and Question 2.6 asks you to track their crystallization paths.

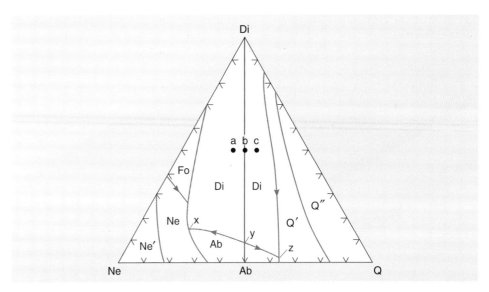

**Figure 2.6** The ternary nepheline(Ne)–silica(Q)–diopside(Di) system at low pressure (0.1 MPa), with Ab representing the intermediate compound albite. Q′ and Q″ represent silica polymorphs that occur instead of quartz at higher temperatures; Ne′ represents a higher-temperature polymorph of nepheline.

**Question 2.6** From Figure 2.6, what will be the sequence of crystallization and final compositions of liquids a, b and c?

What is crucial about Figures 2.5 and 2.6 is that they show how it is possible for quite different (evolved) liquid compositions to evolve and rocks of different mineralogies to crystallize from similar parent compositions (a, b and c). The dividing line Di–Ab plays a key role in this, and is known as a **thermal divide**. It acts as a barrier preventing compositions on one side from crossing to the other side. Of wider importance though, is that this thermal divide also operates in more complex systems (not just in the simple basalt magma analogue represented by Figure 2.6). Even when dealing with real minerals such as clinopyroxene, plagioclase and olivine, those basalt compositions that have an excess of silica (i.e. are silica-oversaturated) evolve to quartz-rich residual liquids, whilst those compositions with insufficient silica (i.e. are silica-undersaturated) evolve to nepheline-rich residual liquids. Only composition b, which lies on the thermal divide, evolves to a feldspar-rich residual liquid with neither normative quartz nor normative nepheline being present.

## 2.3.5 The normative basalt tetrahedron (Ne–Q–Di–Fo)

To add olivine — our fifth and final component — means adding another dimension and creating a quaternary diagram (i.e. a pyramid). We do this by adding normative olivine (as forsterite, Fo) as a fourth apex, to produce a Ne–Q–Di–Fo tetrahedron (Figure 2.7). This is called the **normative basalt tetrahedron** because the normative mineralogy of a basalt can be represented by a point plotted inside the tetrahedron.

You should recall that the dividing line Di–Ab is a thermal divide within the ternary system (Figure 2.6). On Figure 2.7, the Di–Ab thermal divide is the common meeting point of two *planes*, a Di–Ab–Fo plane, and a Di–Ab–En plane. These two planes can be thought of as slicing the tetrahedron into three small sub-tetrahedra, all of which share the Di–Ab dividing line.

And now we can return to why we focused on the three basalt rock names earlier — alkali basalt, olivine tholeiite and quartz tholeiite. Each of these basalt types — defined by their normative mineralogy — lies within one of the three segments of the tetrahedron (Figure 2.7).

The thermal divide between alkali basalts and tholeiitic basalts on Figures 2.5 and 2.6 (Di–Ab) is extended into the basalt tetrahedron (Figure 2.7) as the plane Di–Ab–Fo. This plane is *still* a thermal divide and it *still* separates alkali basalts from

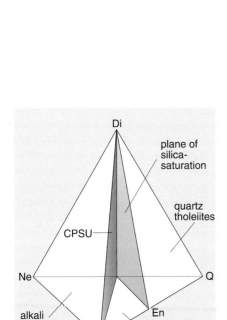

**Figure 2.7** (a) The Di–Fo–Q–Ne tetrahedron, or normative basalt tetrahedron. The 'critical plane of silica-undersaturation' (CPSU), which is the low-pressure thermal divide that separates alkali basalts from tholeiites, is highlighted in brown. A less important 'plane of silica-saturation' — which is not a thermal divide — separates olivine tholeiites from quartz tholeiites (highlighted in green). Note that each of the three basalt types occupies its own separate volume.

tholeiitic basalts, and because of this distinction it is called the **critical plane of silica-undersaturation (CPSU)**. You should note however, that the CPSU only operates at low pressures. With an increase in pressure to above 0.5 GPa (1 GPa = $10^9$ Pa), equivalent to about 15 km depth, the thermal divide breaks down and silica-saturated basalts can evolve to silica-undersaturated liquids. However, this aspect of basalt evolution need not concern you at this stage.

### 2.3.6 Summary of Section 2.3

- Both alkaline and subalkaline basalts erupt in the ocean basins. Ocean ridges are almost exclusively characterized by tholeiites whereas both tholeiites and alkali basalts erupt on large ocean islands. Small ocean islands are dominated by alkali basalts.

- Phase diagrams illustrate the importance of eutectics and thermal divides in controlling the evolution of residual liquid compositions produced during cooling of basalt melt.

- A thermal divide (represented by the CPSU in the normative basalt tetrahedron) acts as a ridge or barrier at low pressures (<0.5 GPa) between residual liquids formed during fractional crystallization of alkali basalts and tholeiites.

## 2.4 Mantle melting

### 2.4.1 Phase diagrams and mantle melting

Basalt magma is derived from the mantle by melting, but to understand how that process operates it is essential to investigate the composition of the mantle itself. Information on the mineralogy, petrology and chemical composition of the mantle is derived from two main sources: (i) samples of peridotite nodules or xenoliths that are occasionally brought to the surface in volcanic eruptions; and (ii) so-called Alpine peridotites and ophiolites found in orogenic belts. These examples show us that the mantle is composed of an ultramafic rock peridotite which has an ultrabasic composition. Ultramafic means that its mineralogy is dominated by ferromagnesian minerals while ultrabasic implies a $SiO_2$ content of <45 wt %. Peridotites can be further subdivided according to their mineralogy and this is described in Box 2.2.

---

**Box 2.2 Mantle rocks and minerals**

The mantle is made up of a rock called peridotite. Peridotite is rich in olivine and contains variable amounts of pyroxene and an aluminous phase. However, it is useful to subdivide peridotite into three types when discussing variations in mantle composition, based on the proportions of olivine to pyroxenes.

Lherzolite. Although dominated by olivine, a lherzolite also contains a low-Ca pyroxene (orthopyroxene) and at least 5% high-Ca pyroxene (clinopyroxene), as well as an aluminium-bearing mineral (garnet, spinel or plagioclase feldspar).

Harzburgite. More olivine-rich than the lherzolite, and contains some orthopyroxene, but clinopyroxene is less than 5% and may be absent. Small amounts of an aluminium-bearing mineral may also be present.

Dunite. This is a rock composed of >90% olivine.

At the surface (atmospheric pressure), lherzolite has a solidus temperature of about 1100 °C, which increases to c. 1500 °C at 100 km depth.

Mantle minerals are members of solid-solution series, and hence are variable in composition and melt over a temperature range. Both olivine and orthopyroxene form solid-solution series between Mg-rich and Fe-rich pure end-members, with mantle types being characteristically Mg-rich.

Clinopyroxenes are more variable, but broadly Ca–Mg-rich. Aluminous phases can be plagioclase, spinel and garnet.

Table 2.4 lists major element analyses of three different peridotites that have been found in the sources described above.

⬤ What can you conclude about the composition of the mantle from the analyses in Table 2.4?

⬤ The mantle cannot be chemically homogeneous.

Table 2.4  Major element analyses of three different peridotites.

| Major element oxides | Lherzolite | Harzburgite | Dunite |
|---|---|---|---|
| $SiO_2$ | 45.35 | 44.69 | 42.57 |
| $TiO_2$ | 0.16 | 0.02 | 0.01 |
| $Al_2O_3$ | 4.26 | 0.86 | 0.19 |
| FeO | 8.24 | 8.17 | 7.76 |
| MnO | 0.14 | 0.12 | 0.12 |
| MgO | 38.17 | 45.04 | 49.19 |
| CaO | 3.39 | 1.09 | 0.14 |
| $Na_2O$ | 0.29 | 0.02 | 0.01 |
| $K_2O$ | 0.03 | 0.01 | 0.00 |
| Total | 100.00 | 100.00 | 100.00 |
| modal mineralogy: | | | |
| olivine | 58 | 73 | 100 |
| orthopyroxene | 25 | 23 | 0 |
| clinopyroxene | 15 | 4 | 0 |
| Al-bearing phase | 2 | 0 | 0 |

Variation in the chemical and mineralogical composition of peridotites provides evidence of chemical variation within the mantle, with some elements varying by up to a factor of 25 (e.g. CaO in the peridotites listed in Table 2.4). But are these peridotites related to each other in any systematic manner? We can use a phase diagram to help us resolve this (Figure 2.8). The dominant minerals in these peridotites are olivine, orthopyroxene and clinopyroxene (Table 2.4), and we can represent these on an Fo–Di–En ternary phase diagram. Olivine is represented by Fo, clinopyroxene by Di, and orthopyroxene by En.

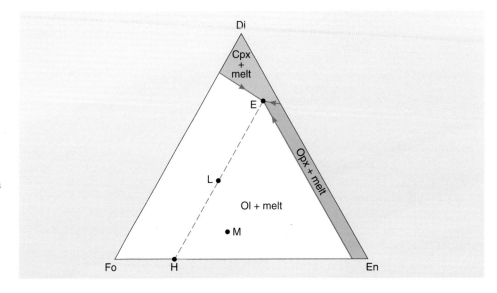

Figure 2.8    Schematic phase diagram of the ternary system Fo–Di–En at 2 GPa (~60 km depth). E is the lowest-temperature eutectic in this system. Arrows on the cotectic curves indicate down-temperature direction. Cpx = clinopyroxene; Opx = orthopyroxene.

So far, phase diagrams have been used to illustrate the effects of fractional crystallization on the composition of a liquid or melt. However, by applying the same principles *in reverse*, a phase diagram can also be used to investigate how a solid rock melts. Consider a lherzolite, L, in Figure 2.8.

● What is the composition of the first melt?

● The first liquid to form has the composition of the eutectic, E.

Remembering that, during crystallization, the last liquid to solidify has a eutectic composition, the first melt to form when a solid is melted also has a eutectic composition. This is an important observation and it shows how any rock with more than one mineral can produce a melt with a very different composition from that of the bulk rock.

● If the peridotite had a composition M, what would be the composition of the first melt?

● Again the melt would have the composition of the eutectic E.

Because peridotite is a rock made up of different minerals, the melt composition is controlled by the location of the eutectic point defined by those minerals, not by the bulk composition of the rock. And so any rock composed predominantly of olivine, enstatite and diopside will produce a eutectic melt. These questions illustrate two important points regarding mantle melting:

1 The composition of a mantle melt is very different from the composition of the mantle.

2 The composition of a mantle melt is broadly independent of the modal mineralogy of the mantle itself.

These two points also explain why basalts have broadly similar abundances of $MgO$, $CaO$, $FeO$, $SiO_2$ and $Al_2O_3$ wherever they occur.

● What happens to L as melting proceeds?

● The production of more melt pushes the composition of L away from the melt composition E towards the Fo–En join along the blue dashed line.

In this simple ternary system, it is the composition of the source rock that changes as a result of the extraction of the eutectic melt.

● What limits the amount of melt that can be produced at E?

● The amount of diopside in L.

The amount of melt that can be produced at E is limited by the amount of diopside in the original composition, L. For example, if lherzolite L contains 15% clinopyroxene and the eutectic composition E contains 60% clinopyroxene, then the amount of melt that can be produced is 15/60 = 0.25 or 25%.

As melting continues, the composition of L migrates closer and closer to the Fo–En join and eventually meets it at H.

● What is the mineralogy of H?

● Fo + En.

H has the mineralogy of a clinopyroxene-free harzburgite. Thus, mantle melting produces basaltic magma which has a composition analogous to a eutectic melt while the residual solid is depleted in clinopyroxene and becomes increasingly more like a harzburgite as melting continues. Figure 2.8 can be regarded as an approximation to an ideal phase diagram for the mantle but a number of important factors are missing.

**Question 2.7**  What other compositional factors should be included in a more representative phase diagram for the mantle?

The effect of adding these extra components to the system is to increase the complexity of the phase diagram to such an extent that it is difficult to represent clearly in two dimensions. However, the addition of iron to the system can be illustrated simply by reference to the Fo–Fa phase diagram, as shown in Figure 2.9. This system is closely analogous to the Ab–An binary solid solution series you explored in Activity 2.2.

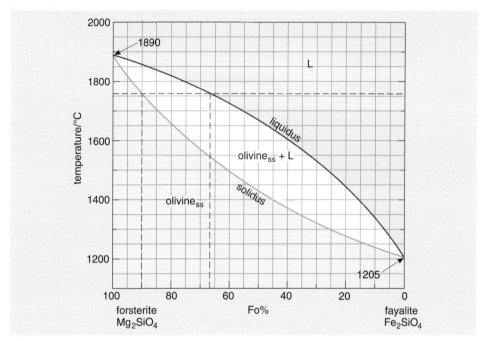

**Figure 2.9**  The phase diagram for olivine, forsterite–fayalite. See text for explanation of dashed lines.

⬤  What is the effect of adding iron to the olivine system?

⬤  It reduces the melting temperature.

Fayalite has a lower melting temperature than forsterite and so most naturally occurring olivines have a lower melting point than pure forsterite.

⬤  Is a melt in equilibrium with olivine richer in iron or poorer than the co-existing solid?

⬤  Richer in iron.

A solid composition with, for example, $Fo_{90}Fa_{10}$ (i.e. 90% forsterite and 10% fayalite) melts at a temperature where that composition intersects the solidus curve. At that temperature, the solid olivine is in equilibrium with a melt that lies on the liquidus curve at the same temperature, as shown by the horizontal dashed line in Figure 2.9. All liquid compositions are more iron-rich than the co-existing equilibrium olivine because pure fayalite has a lower melting point than pure forsterite.

⬤  What does this mean for the MgO : FeO ratio of basalts relative to that in the mantle?

⬤  Basalts will have lower MgO : FeO ratios than the mantle from which they are derived.

This latter point is important and will be developed later.

The addition of an aluminous phase does not move the position of the eutectic significantly although it does reduce its temperature slightly. In summary, the addition of extra components to the melting model increases the complexity of the phase diagram but does not change the major conclusions drawn from the simple system illustrated in Figure 2.8.

● Apart from composition, what other factor has not yet been covered in this discussion of mantle melting?

● Pressure, or depth of melting.

Figure 2.8 illustrates the phase relations of mantle melting at 2 GPa, equivalent to a depth of about 60 km. How applicable is this mineralogy to different depths? Figure 2.10 is a phase diagram of a lherzolite xenolith from a volcanic locality in Arizona. This sample has been the subject of intensive study for many years because its bulk composition closely resembles that from mantle models based on the compositions of meteorites. As well as defining a liquidus and a solidus, the experiments reveal that the mineralogy of the lherzolite varies with pressure. At low pressures, olivine and pyroxenes coexist with plagioclase. At intermediate pressures, plagioclase is replaced by spinel while at high pressures spinel gives way to garnet. In Box 2.2, reference was made to different aluminium-bearing minerals that can be found in peridotites and these distinct mineralogies are simply due to aluminium being located in different phases of increasing density. The transitions from plagioclase lherzolite to spinel lherzolite and from spinel to garnet lherzolite occur in the solid state and without change in the bulk composition of the whole rock: they are therefore examples of metamorphic reactions.

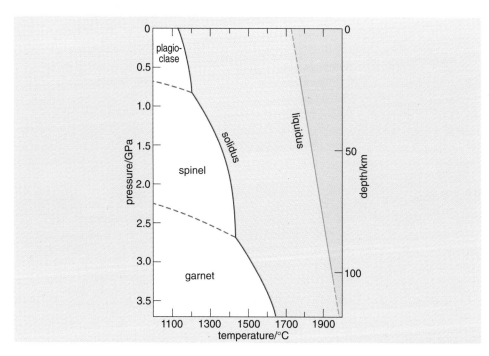

**Figure 2.10** Pressure–temperature phase diagram of lherzolite KLB-1. Yellow regions are solid, green is solid + liquid, and blue is liquid.

## 2.4.2 Decompression melting

From Block 1, we know that heat transfer is controlled by *conduction* in the lithosphere and *convection* in the mantle beneath the lithosphere. Figure 2.11 is a pressure–temperature diagram showing the mantle solidus and a calculated geotherm for typical oceanic crust well away from active plate boundaries. For the mantle to melt, it must cross its solidus — from the (sub-solidus) zone below the solidus into the higher-temperature (partly molten) zone between the solidus and liquidus. However, you can see that the geotherm lies below the solidus, leading to predictions that the mantle is normally solid and contains no melt. So, we need a mechanism that allows the mantle to cross its solidus — and a significant change in pressure and/or temperature is needed.

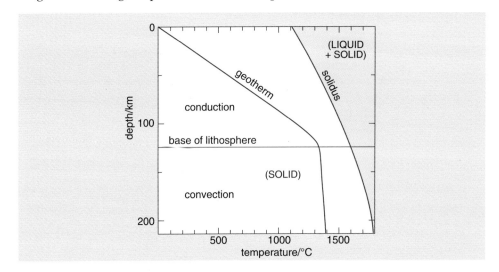

**Figure 2.11** The solidus of mantle peridotite and a calculated geotherm beneath typical oceanic crust situated well away from plate boundaries.

There are two important situations where upward flow of mantle — and the resulting decrease in pressure — allows the mantle to cross its solidus and partially melt. These are: (1) a **mantle plume**, driven by the upwelling of hot, buoyant mantle; and (2) where plate movements stretch and thin the lithosphere, inducing the (passive) upwelling of mantle towards the surface. Figure 2.12 illustrates these two situations, with path x–y–z representing a mantle plume and path a–b–c representing mantle which upwells in response to lithospheric stretching and thinning (see Figure 2.13, which is a sketch showing this, with points a and b shown for comparison with Figure 2.12).

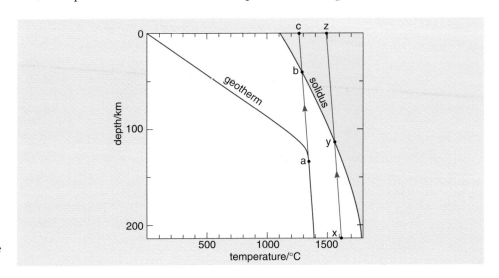

**Figure 2.12** Schematic temperature versus pressure plot showing two paths along which mantle can ascend and cross the solidus, melting and producing basalt magma as it does so. Path a–b–c is broadly analogous to mantle rising beneath spreading ridges, whereas path x–y–z is broadly analogous to the uprise of a hot mantle plume.

**Question 2.8** Compare and contrast the consequences for two packets of mantle rising along the two pathways a–b–c and x–y–z.

At spreading axes, the ocean crust is thinned to just a few kilometres, allowing upwelling mantle to rise to within <10 km of the surface (Figure 2.12). And therefore it is hardly surprising that so much basalt magma is generated at the ~60 000 km of spreading ridges that web the surface of the planet.

However, in regions well away from divergent plate boundaries, the geotherm is mostly below the mantle solidus and so melt is not produced. Only in regions where the mantle geotherm is hotter than normal are the conditions appropriate for melting to occur.

● Can you think of an example where melting occurs away from a plate boundary?

● The best example of so-called **within-plate** magmatism and hence melting is the island of Hawaii.

Hawaii is the most productive basalt volcano on Earth, erupting on average $5 \, m^3$ of basalt every second or about $0.16 \, km^3$ every year. The lithosphere beneath Hawaii is about 60 Ma old and about 75 km thick, so melting must occur below this depth, i.e. at pressures greater than 2.5 GPa. For this to occur the mantle must be hotter than the normal adiabatic gradient (Block 1) and for Hawaii it has been calculated that the mantle must be at least 200 °C hotter than the rest of the convecting mantle.

In summary, on the one hand some basalt magmas signal the location of material upwelling from depth (i.e. mantle plumes), which are independent of tectonic processes — and which result in the mantle crossing its solidus at relatively high temperatures and pressures. On the other hand, there are basalt magmas that signal the passive upwelling of mantle when tectonic forces stretch and thin the crust — and which result in the mantle crossing its solidus at relatively low temperatures and pressures. In the next Section, we will explore how the temperatures and pressures at which the mantle solidus is crossed affect the amount of basalt melt generated.

## 2.4.3 Potential temperature

In this Section, we introduce a concept that is of central importance in modelling melt generation in the mantle and the formation of oceanic crust — **potential temperature**.

Figure 2.14 is a pressure–temperature diagram showing the mantle solidus and liquidus, together with three ascent paths (A, B and C). You can follow the path followed by a packet of mantle that originates at point C, at the same pressure as

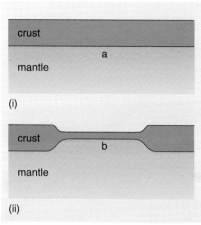

**Figure 2.13** Schematic cross-sections of crust and mantle, showing approximate depths (but not temperatures) of points a and b on Figure 2.12. Cross-section (i) shows crust overlying mantle, whereas (ii) shows crust that has been stretched and thinned, allowing mantle to rise and decompress. Stretching of the crust allows mantle to rise adiabatically, cross its solidus, and melt.

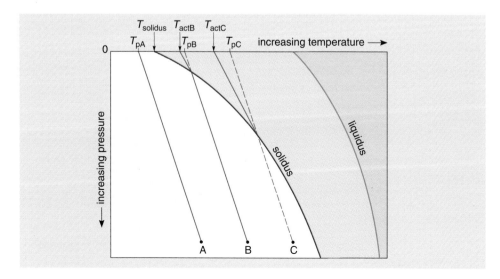

**Figure 2.14** Diagram of pressure against temperature showing the adiabatic ascent paths of mantle at A, B and C, which have identical mantle pressures but different potential temperatures. Note the solidus and liquidus curves of peridotite.

mantle at A and B, but a higher temperature. C rises along the adiabatic temperature gradient, until it meets the solidus. At the solidus, C starts to melt, and continues to rise until it erupts at the surface at temperature $T_{actC}$ (the actual eruption temperature of C).

- Why does the gradient (slope) change once C is above the solidus?

- During partial melting, heat is consumed (latent heat of fusion), and so the mantle+melt packet is cooler, causing the path to be displaced towards lower temperatures.

Had the mantle continued to rise along an adiabatic temperature gradient without melting, it would have reached the surface not at point $T_{actC}$ but at point $T_{pC}$, which is the potential temperature. So, the potential temperature is defined as the *temperature extrapolated from the adiabatic temperature gradient of a particular packet of mantle to the surface.*

Notice that on Figure 2.14 there are three packets of mantle, and so far we have only dealt with C. Consider what would happen if A rose to the surface along an adiabatic path. You should notice that A *never* crosses the solidus, and so this mantle cannot produce any partial melt.

- What does this suggest about the minimum potential temperature of mantle that can partially melt?

- As no partial melt is generated by mantle rising at point $T_{pA}$, the potential temperature of the Earth's mantle must be greater than $T_{pA}$. Furthermore, as the solidus defines the boundary between solid and solid+melt, for mantle to generate any partial melt its potential temperature must be greater than or equal to point $T_{solidus}$, where the mantle solidus intersects the surface.

The implication of this is that in order for the mantle to melt at all its potential temperature must be greater than 1100 °C — the solidus temperature of peridotite at 0.1 MPa (i.e. atmospheric pressure). (*Note:* This statement is only true if the mantle is dry. If there is any water present, this will reduce the solidus considerably, as discussed in Block 3.)

Let us now investigate just how much partial melt is produced, and how this varies with both temperature and pressure. Figure 2.15 is a graph of the melt fraction in lherzolite at temperatures between the solidus and liquidus. Despite these data being obtained at various pressures (0.1 MPa to 1.5 MPa) and on different peridotites, there is a coherent trend.

**Question 2.9** Using Figures 2.14 and 2.15, compare and contrast the behaviour of mantle B and mantle C during adiabatic ascent in terms of: (a) the relative depths (pressures) at which partial melting starts; (b) the maximum degree of partial melting achieved; and (c) the total volume of melt produced.

Your answer to Question 2.9 should have led you to conclude that there is an intimate relationship between potential temperature, the depth at which melting starts and the amount of melt produced.

- If you assume that all of the melt generated during partial melting escapes and forms oceanic crust, what can you say about the relative potential temperatures of mantle that forms thick crust and mantle that forms thin crust?

- The mantle that forms the thicker oceanic crust must have the higher potential temperature.

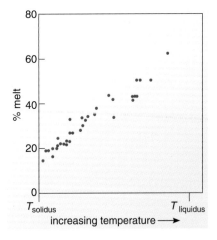

**Figure 2.15** Graph of melt fraction in lherzolite at temperatures between the solidus and the liquidus, with data collected from various pressures (0.1 MPa to 1.5 MPa) and various lherzolite samples.

The answer to the above question assumes of course that other factors, such as spreading rate, do not have an effect on melting or are constant. However, for most moderate to fast spreading axes (> 1 cm yr$^{-1}$ half-spreading rate), the relationship holds true.

Turning this relationship around reveals the possibility that measurements of ocean crust thickness can be used to provide information on mantle potential temperature. So, simply by assuming that the thickness of ocean crust equals the melt thickness, we can get a picture of potential temperature variations within the mantle. This in turn allows us to investigate the thermal structure of the mantle.

It is possible to apply the concept of potential temperature further. Look at Figure 2.16, which consists of three diagrams encapsulating the results from a series of calculations showing how potential temperature is related to: (a) maximum melt thickness (i.e. the thickness of the ocean crust); (b) melt fraction (i.e. the amount of melting); and (c) depth of melting. To gain an appreciation of how these diagrams reveal important information about melting processes, you should now attempt Question 2.10.

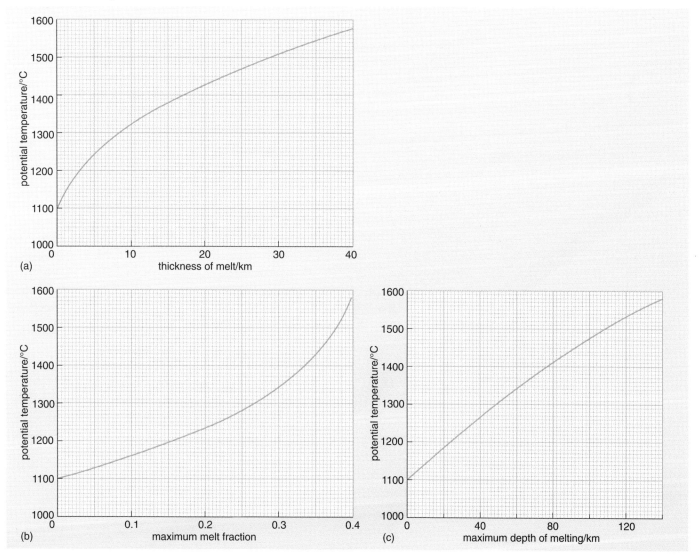

**Figure 2.16** Calculated relationships between potential temperature of the mantle and: (a) the thickness of melt produced and extracted; (b) maximum mass fraction of partial melt in the melting zone; and (c) maximum depth at which melting of adiabatically ascending mantle will start to partially melt. These graphs assume decompression continues to atmospheric pressure, 0.1 MPa.

**Question 2.10**   (a) Given an average ocean crust thickness of 7 km, estimate the potential temperature of the underlying mantle. (b) From the potential temperature estimated in (a), estimate the maximum melt fraction to which this corresponds. (c) Using the same potential temperature you estimated in (a), estimate the maximum depth at which partial melting would begin, and using Figure 2.10, state what type of lherzolite is stable at this depth.

From our considerations so far, we can draw two important conclusions. First, from Block 1 you should recall that (away from spreading ridges) most of the ocean crust ranges from 5 km to 10 km in thickness, with an average thickness of 7 km. If you apply the methodology used to gain the correct potential temperature in Question 2.10 to melt thicknesses of 5 km and 10 km, then you would discover that the 5 km difference in ocean crust thickness stems from only an 80 °C range in potential temperature.

●   What does this suggest about the thermal structure of the upper mantle?

●   For such a small potential temperature variation to exist, the upper mantle must be kept well stirred by solid-state convection.

Secondly, we have established that higher proportions of melt and greater volumes of melt are generated by adiabatically rising mantle crossing its solidus at greater pressures and temperatures. Thus, by studying basalts (age, area, volume, eruption rate, composition, and so on), Earth scientists can reconstruct the conditions in the mantle that led to their formation and eruption.

## 2.4.4   Summary of Section 2.4

* Although the mantle undergoes solid-state convection, variations in peridotite composition suggest that it is not chemically homogeneous.

* Lherzolite is the dominant upper mantle composition, and when it partially melts it produces a liquid (i.e. basalt melt) and leaves a solid residue depleted in clinopyroxene.

* The upper mantle is mineralogically layered, and with increasing depth (pressure) plagioclase lherzolite, then spinel lherzolite, then garnet lherzolite are encountered. This change is brought about by the pressure-dependent stability ranges of the Al-bearing phases (plagioclase, spinel and garnet).

* The mantle partially melts when it moves from the sub-solidus field and crosses the solidus into the solid + melt field that lies between the solidus and liquidus curves.

* Upwelling of mantle under adiabatic conditions results in decompression melting, two important examples of which are mantle plumes and lithospheric thinning. Mantle plumes are independent of tectonic influence, whereas lithospheric thinning is intimately linked to global tectonics.

* The potential temperature of the mantle is the temperature it would have if moved to the surface along an adiabatic temperature gradient.

* Assuming that melt thickness equals ocean crust thickness allows relationships between potential temperature, the amount of melt generated, and the depth of melting to be established.

* The small potential temperature difference (c. 80 °C) between typical ocean crust that ranges from 5 km to 10 km in thickness suggests that the upper mantle is kept well stirred by convection.

## 2.5 Basalts — origins

So far you have seen how the mantle melts, and that the liquids formed during partial melting of a typical lherzolite are basalt-like in composition. We refer to the liquids formed in the mantle during partial melting as **primary melts** (or primary magmas), but does this mean that all basalts are primary melts? The simple answer is no, as basalts can be modified before they reach the surface — for instance by fractional crystallization (Section 2.3). So it would be helpful to have a simple method that would enable us to discriminate between basalts that could represent primary melts from those that have been modified en route to the surface. We need to develop some criteria by which primary basalts can be recognized.

### 2.5.1 Identifying a primary basalt

The compositions of partial melts produced in the mantle must be closely related to mantle mineralogy, and the dominant mineral in the mantle is olivine. Therefore, assuming that melts are produced in equilibrium with residual mantle, the MgO and FeO contents of the liquid will be controlled by the composition of the residual olivine. If you know the composition of mantle olivine, then the composition of a co-existing liquid can be determined from the olivine phase diagram.

You should recall from Section 2.4 that the mantle is MgO-rich. Analyses of a wide range of mantle samples reveal that olivines in mantle peridotites are also MgO-rich and they have a restricted range of compositions, from $Fo_{92}$ to $Fo_{88}$. By referring to the olivine phase diagram (Figure 2.9), you should be able to answer the following question.

> **Question 2.11**  What is the range of liquid compositions that would be generated by melting mantle olivines in the compositional range $Fo_{88}$ to $Fo_{92}$?

Your answer to Question 2.11 should have confirmed the prediction made above that the liquids are richer in fayalite (i.e. more iron-rich) than the mantle itself. You should also have noted that the liquids represent a narrow range of Fo contents.

How can we relate this result to real basalts? The forsterite content of the liquid in equilibrium with mantle olivines can also be expressed as the atomic ratio of $Mg^{2+}$ ions in the liquid to the total of $Mg^{2+} + Fe^{2+}$ and is known as the **magnesium number — Mg#**. This parameter is easily calculated for any composition and the method is summarized in Box 2.5. In general, basalts with Mg# >65 can be considered primary.

> **Question 2.12**  Calculate the Mg# of the three basalt analyses in Table 2.3, and comment on the likelihood that they represent primary melts.

The Mg# is a very useful criterion for recognizing basalts that may represent primary magmas derived directly from their mantle source region. But, as with all geochemical discriminants, the Mg# has to be used with some caution.

● What is the first mineral to crystallize from most basalts at low pressure?

● Olivine.

● How will removal of early-formed olivine affect the Mg# of the basalt liquid?

● It will reduce it.

Again, this effect can be explored using the olivine phase diagram in Figure 2.9. Consider a basalt magma with a Mg# of 70.

● What is the forsterite content of an olivine in equilibrium with this magma?

● From Figure 2.9, the olivine has a composition of $Fo_{91}$.

In as much as the liquid in equilibrium with a solid has a lower Mg# than the solid, so crystals separating from a melt will have high Mg#. Removal of even small amounts of olivine of this composition will push the liquid to lower values of Mg# and because forsteritic olivine is rich in MgO, the liquid will become relatively depleted in MgO also. Continuing this process produces basaltic magma with lower and lower Mg# and MgO contents. Such magmas are generally described as **evolved magmas** because they have undergone some fractional crystallization since separation from their source regions, and the rocks that crystallize from these magmas when they erupt are often described as evolved rocks.

What if an evolved magma now incorporates some olivine crystals that formed from an earlier batch of magma? Many basalts include aggregates of olivine crystals and broken olivines that appear to have been inherited from another source and not crystallized directly from their host rock. Such crystals are known as **xenocrysts**. They will clearly increase the Mg# of the bulk composition, possibly to a value that appears primitive. The message here is to treat Mg# information with caution and always relate it to the texture of the rock — if there is evidence for olivine xenocrysts, then the Mg# of the bulk rock may be misleading.

### Box 2.5  Mg number, Mg#

The Mg number of a basalt (or any other rock) is defined as the ratio $100\,Mg/(Mg+Fe^{2+})$ in which Mg and $Fe^{2+}$ are expressed in atomic proportions. The atomic proportions are calculated from the wt % oxides by dividing the wt % oxide by its relative molecular mass. The relative molecular mass of MgO is 40 (i.e. the sum of the relative atomic masses of Mg and O: 24 + 16 = 40) and the relative molecular mass of FeO is 72 (i.e. the sum of the relative atomic masses of Fe and O: 56 + 16 = 72). A complete rock analysis will include Fe as both FeO and $Fe_2O_3$, but it is the $Fe^{2+}$ as FeO that we use in calculating the Mg# (i.e. $Fe^{3+}$ is not used).

For example, a basalt containing 12 wt % MgO and 8 wt % FeO will have an Mg# of:
$100 \times (12/40)/[(12/40) + (8/72)] = 73$.

The Fo value of an olivine (or its liquid) is analogous to the Mg# of a rock — as the Fo value of an olivine is equivalent to the atomic proportions $100\,Mg/(Mg + Fe^{2+})$. Thus, we have a means of directly comparing the chemical compositions of basalt magmas and melts in equilibrium with mantle olivines.

The Mg# is a useful tool that allows us to comment on the likelihood of whether a particular basalt is a primary melt, but MgO and FeO are only two of the ten or so major elements of basaltic rocks. The next stage is to investigate how other aspects of the compositions of primary basalts vary with physical conditions of melting in the mantle.

## 2.5.2 High-pressure experimental work

In addition to carrying out experiments at atmospheric pressure to investigate the effects of fractional crystallization on basaltic magmas, experimental petrologists have also conducted experiments at higher pressures to simulate conditions in the deep crust or even in the upper mantle. So far you have seen how the peridotite solidus varies with pressure but detailed high-pressure experiments have also shown that the compositions of liquids produced vary with pressure and temperature.

An experimental technique developed during the early 1990s has generated high-quality data on the high-pressure partial melting of the mantle. The diamond aggregate method involves using a graphite capsule within which a clump of diamond aggregate is surrounded by powdered peridotite. As diamond is strong and does not deform under the pressures used in the experiments, the diamond aggregate retains its pore space. During the experiment, the melt migrates into the diamond pore space where it is isolated from the residual peridotite. After a period of up to 24 hours, the sample is rapidly cooled (quenched), and the major element composition of the partial melt is determined using an electron microprobe.

● Why does the melt have to be cooled rapidly (quenched)?

● To inhibit formation of crystals, and back reaction with residual peridotite.

Table 2.5 lists the starting material and summary results of one such series of experiments. The starting material was a peridotite KLB-1, the spinel lherzolite xenolith from Kilbourne Hole in Arizona used to construct the phase diagram in Figure 2.10. This is a typical mantle peridotite with a modal mineralogy of 58% olivine, 25% orthopyroxene, 15% clinopyroxene and 2% spinel.

● From what range of depths in the mantle was this sample derived?

● The peridotite phase diagram (Figure 2.10) shows that spinel is stable from 0.9 to 2.6 GPa equivalent to a depth range of 30–90 km.

**Table 2.5**  Analyses of spinel lherzolite KLB-1 and partial melts produced at different pressures. Note that all iron is reported as FeO (total).

| Major element oxides | KLB-1 | 1 GPa 1300 °C | 1 GPa 1400 °C | 1.5 GPa 1400 °C | 2.0 GPa 1425 °C | 2.5 GPa 1425 °C | 3.0 GPa 1500 °C |
|---|---|---|---|---|---|---|---|
| $SiO_2$ | 44.48 | 51.32 | 51.59 | 49.88 | 48.74 | 47.97 | 45.67 |
| $Al_2O_3$ | 3.59 | 19.09 | 12.58 | 13.78 | 13.16 | 14.88 | 14.33 |
| FeO (total) | 8.10 | 6.38 | 7.95 | 7.92 | 8.80 | 9.43 | 9.59 |
| MgO | 39.22 | 8.14 | 16.41 | 15.74 | 15.69 | 13.36 | 16.73 |
| CaO | 3.44 | 8.85 | 9.42 | 10.69 | 11.06 | 10.23 | 10.64 |
| $Na_2O$ | 0.30 | 4.60 | 0.91 | 1.04 | 1.37 | 2.37 | 1.80 |
| $K_2O$ | 0.02 | 0.27 | 0.05 | 0.04 | 0.13 | 0.82 | 0.07 |

● Do the compositions of the experimentally derived melts look like basalts?

● Yes, they do. The analyses in Table 2.5 are very similar to the compositions of the basalts in Table 2.3. They generally contain between 45–52% $SiO_2$ and <5% total alkalis.

If you scan the data in Table 2.5, and compare these compositions with the three basalt analyses in Table 2.3, you should be able to see that the melts generated during the experimental runs on KLB-1 are definitely basalt-like in chemical composition.

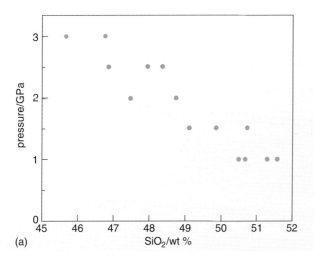

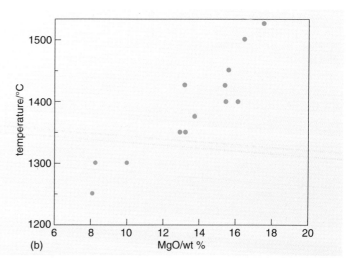

**Figure 2.17** (a) $SiO_2$ contents in melts formed by partially melting spinel lherzolite KLB-1, plotted against pressure. (These data were obtained at various temperatures.) (b) MgO contents in melts formed by partially melting spinel lherzolite KLB-1, plotted against temperature. (These data were obtained at various pressures.)

● Do the experimental melts have the same compositions?

● No. $SiO_2$ for example varies from 51.6% to 45.7%.

You may also have noted that $SiO_2$ decreases as pressure increases. This is illustrated in Figure 2.17, which includes the results from the complete experimental data set. Other compositional parameters also vary with pressure and temperature. In particular MgO shows a reasonable correlation with temperature, increasing in the melt as temperature increases. This is shown in Figure 2.17b.

These experimental observations are very important because they allow some constraints to be placed on the physical conditions of melting within the mantle. One important point to note is that the MgO content of a melt generated at a given temperature is insensitive to changes in pressure. Likewise, the $SiO_2$ content at a given pressure is insensitive to temperature. So, these two compositional parameters can be used with some confidence as indicators of the conditions of melting over the ranges of temperature and pressure covered in the experiments. For example, given two basalts with high Mg numbers (indicating that they could be primary melts) and different $SiO_2$ contents, you might conclude that the basalt with the lower $SiO_2$ content was generated at a higher pressure.

> **Question 2.13**  Can you think of a possible limitation on the use of this approach to determining melting conditions in the mantle?

● If two primary basalts have similar $SiO_2$ contents but different MgO contents, what can be deduced about the relative temperatures and pressures during partial melting to generate the two magmas?

● The similar $SiO_2$ content of the two basalts would suggest that they were generated at similar pressures, but the different MgO contents would suggest that the basalt with the higher MgO content represents a primary melt generated under higher temperature conditions.

> **Question 2.14**  (a) Given that basalt 2 in Table 2.3 could represent a primary melt, estimate the pressure and temperature at which partial melting took place. (b) State which type of lherzolite would probably have been stable at these temperatures and pressures.

In previous Sections, we continually emphasized the differences between alkali basalts and tholeiites. We concluded that ocean island basalts (OIB) are produced when mantle plumes trigger decompression melting of the mantle, and that tholeiites erupted at mid-ocean ridges (MORB) are produced via shallow (passive) decompression melting. You also saw that partial melting in mantle plumes occurs at higher pressures and temperatures compared with melting beneath oceanic spreading ridges.

● Given the higher temperatures and pressures at which the OIB primary melts are generated, what differences would you predict in the $SiO_2$ and MgO contents of an OIB and a MORB?

● The higher pressure of the OIB should result in lower $SiO_2$, whilst the higher temperature should result in higher MgO (both relative to the MORB).

Look now at Table 2.6, which contains partial analyses of a MORB (basalt A) and an OIB (basalt B). Both of these basalts could represent primary melts because their Mg# fall within or above the 65–73 range for primary melts. When comparing these two partial analyses, it is clear that the OIB has lower $SiO_2$ (indicating higher pressures), and also higher MgO (indicating higher temperatures). However, before you draw hard and fast conclusions from this rather simple analysis, you should note that while the Mg# of the MORB is within the predicted range of primary melts, it is lower than the Mg# of the OIB. It is therefore possible that the MORB has lost some olivine and its MgO content is lower than that of the true primary magma. The MgO content of the primary magma can be calculated by incrementally adding olivine of the equilibrium composition until the added olivine reaches the composition of $Fo_{90}$. The details of that procedure are not covered in this Course but they reveal that primary MORB seldom have MgO contents greater than 12 wt %. This value is still lower than that of basalt B and so the conclusion that MORB is generated at a lower temperature than OIB is generally correct.

Table 2.6   Analyses of three basalts, together with Mg# and all iron recalculated as FeO (total). Basalt A is a typical olivine tholeiite from an ocean ridge environment, whilst basalts B and C are from the island of Hawaii.

|  | Basalt A MORB | Basalt B OIB | Basalt C alkaline OIB |
|---|---|---|---|
| $SiO_2$ | 49.4 | 47.8 | 46.4 |
| $TiO_2$ | 0.84 | 1.80 | 2.23 |
| $Al_2O_3$ | 16.3 | 11.2 | 14.2 |
| $Fe_2O_3$ | 1.48 | 1.92 | 4.09 |
| FeO | 7.52 | 9.78 | 8.91 |
| MnO | 0.24 | 0.19 | 0.19 |
| MgO | 9.50 | 16.0 | 9.47 |
| CaO | 12.4 | 8.63 | 10.3 |
| $Na_2O$ | 2.15 | 1.80 | 2.85 |
| $K_2O$ | 0.10 | 0.35 | 0.93 |
| Mg# | 69.0 | 75.0 | 66.0 |
| FeO (total) | 8.85 | 11.5 | 12.6 |

Question 2.15   Basalt C is an alkali basalt also from Hawaii. (a) What pressure and temperature conditions are indicated by its $SiO_2$ and MgO contents? (b) How do these conditions compare with those for the tholeiite, basalt B?

As with the discussion of the MgO content of the MORB (basalt A), there are considerable problems in assessing the temperature at which basalt C was generated because of the unknown effects of olivine fractionation.

The important aspect of this exercise is that both tholeiitic and alkaline basalts have their own primary melts that are generated under different pressure and temperature conditions. You should recall from Section 2.2 that both alkali and tholeiitic basalts are found on Hawaii and both basalt B and C are from Hawaii. Their compositional differences suggest different conditions for melting in the mantle beneath the same volcanic island.

### 2.5.3  Summary of Section 2.5

- The Mg# of a basalt is analogous to the Fo value of an olivine (or its liquid), and provides a first-order means of assessing whether or not a basalt could represent a primary melt.

- Mantle olivines have a restricted range of composition ($Fo_{88-92}$), and the range of liquids (i.e. primary melts) in equilibrium with them is also restricted (Mg# = 65–73).

- High-pressure experimental work on spinel lherzolite KLB-1 indicates the effect that pressure and temperature have on the compositions of primary melts.

- $SiO_2$ is strongly controlled by pressure, whereas MgO is strongly controlled by temperature.

- Alkali and tholeiitic basalts have their own primary melts.

## 2.6   Trace elements in basalts

In the previous Sections, you have investigated the differences in the major element compositions of different basaltic magmas, and how they reflect the effects of both fractional crystallization and the pressures and temperatures of melting. In this Section, you will explore how information from trace elements can provide further information on the depth and degree of melting and also reveal differences in the source regions of different basalts.

### 2.6.1  Compatible and incompatible elements

A trace element is one that is not an essential component of a major silicate mineral phase. Trace elements are generally presented as concentrations of the element alone expressed as parts per million (ppm), usually in concentrations below 1000 ppm. Their distribution between minerals and a melt phase is described with reference to a parameter known as a **partition coefficient**, denoted $K_d$ for a single mineral or $D$, the **bulk distribution coefficient** for a rock made up of different minerals. (The use of the terms partition and distribution in this context are often interchanged.) $K_d$ is defined as the concentration of an element i in a mineral divided by its concentration in a coexisting liquid. More formally the $K_d$ for element i is given by:

$$K_d{}^i = c^i{}_m / c^i{}_l$$

where $c^i{}_m$ is the concentration of element i in the mineral and $c^i{}_l$ is the concentration of element i in the liquid or melt.

Clearly, $K_d$ can have any value but an important distinction is between elements that fit easily into a mineral structure and those that do not. An element that fits into a mineral and therefore has a higher concentration in the mineral than the melt is said to be **compatible** ($K_d > 1$). An element that does not fit into a mineral structure, by contrast, and that has a low $K_d$ is said to be **incompatible** ($K_d < 1$). Examples of trace elements and partition coefficients are shown in Table 2.7. The $D$-value of a rock is the weighted sum of the individual partition coefficients, namely:

$$D = x^i K_d{}^i + x^j K_d{}^j + \ldots$$

for a rock composed of minerals i, j,… in the modal proportions $x^i$, $x^j$, …

**Table 2.7** Values of partition coefficients for selected trace elements between mantle minerals and basaltic liquids.

| Element | Olivine | Clinopyroxene | Orthopyroxene | Garnet |
| --- | --- | --- | --- | --- |
| Rb | 0.000 03 | 0.000 4 | 0.000 02 | 0.000 02 |
| Ba | 0.000 03 | 0.000 3 | 0.000 03 | 0.000 03 |
| Nb | 0.000 05 | 0.009 | 0.003 | 0.004 |
| La | 0.000 2 | 0.054 | 0.003 | 0.000 7 |
| Sm | 0.000 9 | 0.27 | 0.004 | 0.22 |
| Zr | 0.001 | 0.26 | 0.012 | 0.20 |
| Y | 0.008 | 0.47 | 0.015 | 2.00 |
| Yb | 0.024 | 0.43 | 0.038 | 4.00 |

Trace elements can be used to investigate any magmatic process in which minerals and liquids are involved, but in this case we are interested in the effects of melting and how different basaltic melts have different trace element abundances and particularly incompatible element abundances.

## 2.6.2 The partial melting equation

The concentration of a trace element in a partial melt can be evaluated using the following equation:

$$C_l = C_0/(D + F(1-D)) \tag{2.1}$$

where: $C_l$ = concentration of a trace element in the liquid (partial melt);

$C_0$ = concentration of a trace element in the solid (e.g. peridotite) before melting;

$D$ = the bulk distribution coefficient for the trace element between the solid and melt;

$F$ = the degree of partial melting expressed as a melt fraction between 0 and 1 (for a 10% melt, $F = 0.1$).

This equation may be applied to trace elements and low abundance major elements such as potassium ($K_2O$), titanium ($TiO_2$) and phosphorus ($P_2O_5$) that do not easily fit into the common mantle silicate minerals and so behave as incompatible elements.

> **Question 2.16**  What would be the effect on the $K_2O$ content of a basalt melt (i.e. $C_l$) if (a) $C_0$, (b) $D$ and (c) $F$ were increased?

> **Question 2.17**  Given that alkali basalts tend to have higher concentrations of alkali elements (including $K_2O$) than tholeiites, what does this tell us about the conditions of formation of these two basalt types?

From this simple reasoning, you should have deduced that high concentrations of alkali elements in the mantle source, low degrees of melting and the presence of residual minerals with low $K_d$-values for $K_2O$ favour the formation of alkali basalts. Conversely, tholeiites may result from partial melting of an alkali-poor mantle by higher degrees of melting, possibly with residual minerals with high partition coefficients for $K_2O$. These different possibilities are explored in the following Section.

### 2.6.3  Trace elements in MORB and OIB

The concentrations of incompatible elements vary by many orders of magnitude in basaltic rocks and simple lists of element concentrations do not convey information about rocks in an easily digested manner. Consequently, igneous geochemists have sought various ways of displaying trace element information and one of the more popular methods is to plot abundances normalized to concentrations in some reference material. Reference materials vary from chondritic meteorites to estimates of the primitive mantle and even average MORB. The most commonly used normalizing values are those of primitive mantle which is calculated from the trace element abundances in meteorites and is an estimate of the composition of the mantle before the extraction of the continental crust. Elements are displayed so that the most incompatible is on the left and the most compatible on the right. Such a diagram is shown in Figure 2.18, which displays trace element abundances in a MORB from the Mid-Atlantic Ridge. You should note two aspects of this profile. First, the data define a smooth trend in which the most incompatible elements have the lowest normalized concentrations while the most compatible have higher normalized concentrations. Secondly, the abundances of the most compatible elements are 5–10 times those of primitive mantle whereas the most incompatible elements have lower abundances.

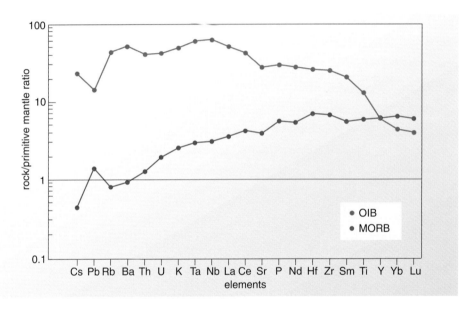

**Figure 2.18** Abundances of trace elements in a sample of MORB and an average OIB, normalized to element concentrations in the primitive mantle.

Does this depletion reflect a source characteristic, i.e. $C_0$ and/or $D$, or the melting process i.e. $F$ and/or $D$?

You have deduced above that low abundances can be produced by either low source concentrations or relatively high $D$-values. However, the elements are arranged so that the most incompatible elements are at the left of the diagram. So their low abundances cannot be controlled by $D$-value and must be a reflection of the composition of their source region. We can now make a highly significant observation: **the source region of MORB is depleted in incompatible trace elements** with respect to the composition of primitive mantle. This aspect of basalt geochemistry is explored further in the following Activity.

### Activity 2.3   Using the partial melting equation to explore MORB trace elements

This Activity involves some simple calculations using the partial melting equation to draw conclusions about the trace element geochemistry of mid-ocean ridge basalts (MORB), and may take about two hours to complete.

The above Activity shows that the source region of MORB is depleted in the most incompatible trace elements relative to primitive mantle. It also shows that while the concentration of an incompatible element in a melt is strongly dependent on the amount of melting, the ratio of one incompatible trace element to another is not strongly affected, unless the melt fraction is similar in magnitude to the $D$-value of the least incompatible element. Thus, partial melting does not affect ratios such as La/Nb, La/Ba and Th/Ta, which have similar values in basalts as in their source regions. This is an important conclusion and allows geochemists to investigate compositional variations in the mantle through the trace element composition of basalts.

### 2.6.4  Trace elements in OIB

The incompatible element abundances of an average ocean island basalt are also plotted in Figure 2.18. The OIB profile is different from that of MORB.

● Compare the two analyses in Figure 2.18. In what way are the mantle-normalized profiles different?

● The OIB has higher concentrations of the most incompatible elements but lower abundances of the heavy rare earth elements, Yb and Lu.

Why should this be? Returning to the results of Activity 2.3, you should have noticed that as the melt fraction decreases, the abundance of the most incompatible elements increases. Therefore, higher incompatible element abundances in OIB suggest smaller melt fractions.

● What other compositional feature of OIB also reflects smaller melt fractions?

● You should recall from Section 2.5.2 that OIB have higher alkali contents than MORB and these were interpreted as the result of smaller melt fractions.

This conclusion assumes that both MORB and OIB are derived from source regions with the same composition, similar alkali element contents and similar incompatible element abundances.

● How can we test this assumption?

● Compare the trace element ratios of MORB and OIB.

If MORB and OIB are derived from similar source regions, then they should have the same ratios of the most incompatible ratios. Figure 2.19 shows a plot of the Ba/Nb and La/Nb ratios in selected MORB and OIB and this reveals that they

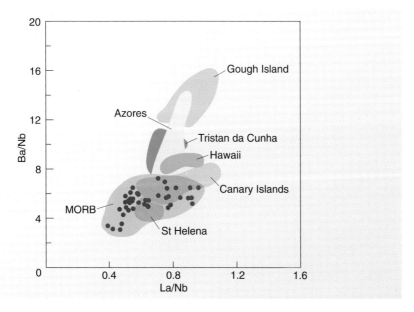

**Figure 2.19**  A plot of La/Nb and Ba/Nb for some MORB and selected OIB. Most of the islands are from the Atlantic, except of course for Hawaii. The MORB data are from both the Mid-Atlantic Ridge and the East Pacific Rise.

are subtly different. MORB has lower Ba/Nb ratios than many OIB, although islands such as St Helena and the Canaries do show some overlap.

The observation that despite the depletion in incompatible elements MORB still has a smooth abundance profile implies that the element depletion is itself controlled by partition coefficients and hence by solid-melt partitioning. The consensus view now is that this depletion in the worldwide source region of MORB is the result of the extraction of the continental crust from the mantle over geological time. While the details of this process are beyond the scope of this Course, you should appreciate that the MORB and OIB source regions are compositionally distinct.

The second difference between MORB and OIB concerned the abundances of Yb and Lu, which are higher in MORB than OIB.

●   What might be the possible reasons for this difference between MORB and OIB?

●   The $D$-value for the heavy rare earth elements (HREE) might be higher during the generation of OIB or the OIB source region may be depleted in these elements. (We have already ascertained that the melt fraction is smaller during the generation of alkali basalts, so varying $F$ cannot of itself produce this difference.)

●   Which mineral accommodates the HREE most easily?

●   Table 2.7 shows that Yb is compatible in garnet.

●   At what depths are OIB produced?

●   >60 km (see Section 2.5).

●   What is the mineralogy of the mantle at these depths?

●   Garnet peridotite (see Figure 2.10).

So the depletion of the HREE in OIB is consistent with melting in the presence of residual garnet. But this observation does not preclude a further difference in the composition of the mantle source region of OIB. Distinguishing between these two possibilities (i.e. mineralogy versus low source abundance) is not simple without recourse to further information concerning the known variation of the HREE in the mantle. Analyses of these elements in mantle peridotites reveal that their abundances do not vary as much as those of the more incompatible elements La, Rb, Ba, Th etc. but appear to remain close to estimates of their abundances in primitive mantle. This suggests that variations in the HREE abundances in OIB are largely a product of varying amounts of residual garnet in their source regions and not the product of variable source compositions. As such, these elements provide further critical evidence that can be used in conjunction with the major elements, to define the conditions under which a basaltic melt was generated.

## 2.6.5   Summary of Section 2.6

*   The distribution of trace elements between solid and melt is described by partition coefficients.

*   Compatible elements have partition coefficients >1; incompatible elements have partition coefficients <1.

*   High incompatible trace element concentrations are favoured by small melt fractions, low $D$-values and high source concentrations.

*   Alkali basalts are the product of smaller melt fractions than are tholeiites.

- The source region of MORB is depleted in incompatible trace elements.

- Low HREE concentrations in OIB result from the presence of residual garnet during melting.

- The source region of OIB has different incompatible element ratios than the source of MORB.

## Activity 2.4   Basalts from Iceland

Activity 2.4 explores the generation of basalts in Iceland and the implication for the thermal structure of the underlying mantle. Allow about four hours for this Activity.

## Objectives for Section 2

Now that you have completed this Section, you should be able to:

2.1   Understand all the terms printed in **bold**.

2.2   Classify a basalt according to its modal and normative mineralogy and bulk composition.

2.3   Describe and identify the mineralogy and texture of a basaltic rock in thin section.

2.4   Understand the reasons for the differences between alkali and tholeiitic basalts with reference to relevant phase diagrams.

2.5   Describe the path of a crystallizing liquid on both binary and ternary phase diagrams.

2.6   Use relevant phase diagrams to describe how the mantle melts and discuss the effects of solid solutions and eutectic points.

2.7   Explain the effects of melting on the mineralogy of the mantle.

2.8   Define what is meant by the term decompression melting and potential temperature and how mantle potential temperature affects the amount of melt produced, the maximum melt fraction and the maximum depth of melting.

2.9   Calculate the Mg# of a basalt and use that to define the composition of a primary melt.

2.10  Deduce conditions of melting from variations in the major element composition of primary magmas.

2.11  Calculate the trace element concentration in a melt using the partial melting equation.

2.12  Describe the differences between trace element abundances in MORB and OIB and discuss the causes of those differences.

# 3 The Kenya Rift

## 3.1 Introduction

In Section 2, you studied how basaltic magma is generated beneath mid-ocean ridges by the upwelling of mantle peridotite in response to lithospheric thinning, whereas beneath ocean islands magma is produced from the active upwelling of hot mantle from depth. The oceanic system is relatively simple because the oceanic lithosphere has a simple structure. Block 1 emphasized the differences between the oceanic and continental lithosphere, showing that continental lithosphere is older and thicker with a more variable composition and a much less uniform structure. Consequently, the connections between extensional tectonics, the thermal state of the underlying mantle and surface magmatism in continental regions are more complex. In this Section, we move on to a detailed examination of one of the best examples of present-day continental rifting — the East African Rift — to investigate these links further and discover why continents rift and drift apart.

You are probably familiar with the East African Rift system from natural history films and TV programmes. Its best-known part is the Kenya or Gregory Rift but this is only one part of a continental-scale feature that stretches from the Red Sea in the north, through Ethiopia, Kenya, Uganda, Tanzania to Malawi in the south, through more than 30° of latitude or over 3000 km length. It is the most prominent topographic feature of continental Africa and hints at geological processes operating on a large scale. But just what are those processes? In a geological sense, what is happening within and beneath the East African Rift? It is the aim of this Section to increase your understanding of the processes that have lead to the development of the East African Rift and its significance for the initial stages of continental break-up.

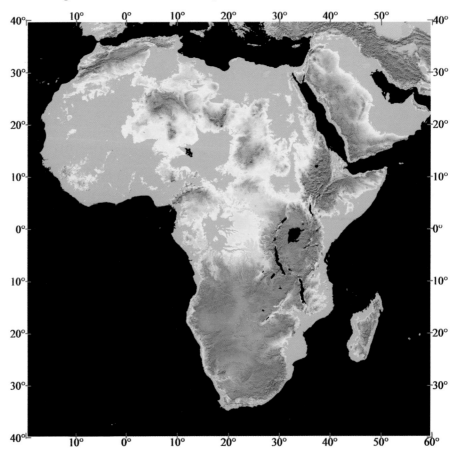

**Figure 3.1** A digital elevation model (DEM) of Africa. Lowest altitudes in green (<1000 m), followed by yellow (1000–2000 m), brown (2000–3000 m), and grey (>3000 m).

## 3.2 The geography of the East African Rift system

Figure 3.1 is a digital elevation model of Africa on which the East African Rift can be seen, cutting across the eastern side of the continent. The rift is characterized by extreme contrasts in elevation, spanning heights in excess of 4000 m in the rift flanks to places on the rift valley floor that are below sea-level. These topographic extremes contrast with the rest of Africa which is characterized by broad swells and extensive plains. Apart from the Atlas mountains in Morocco, there are no dramatic linear mountain belts, such as the Alps, Himalayas and the Andes.

● Can you suggest a reason why this is the case?

● With the exception of northern Africa, there are no destructive plate margins around the edges of the continent.

Africa, like Australia and Antarctica, does not appear to be affected by convergent plate motions, hence the lack of linear fold mountain belts. Despite this apparent lack of plate-scale activity, the plate-scale tectonics of Africa are very important to the evolution of the rift and we shall return to them in a later Section.

Look at Figure 3.2, a close-up of the whole rift system.

● What can you say about the mean elevation of the area affected by the East African Rift?

● Much of the rift is concentrated in regions of high average elevation.

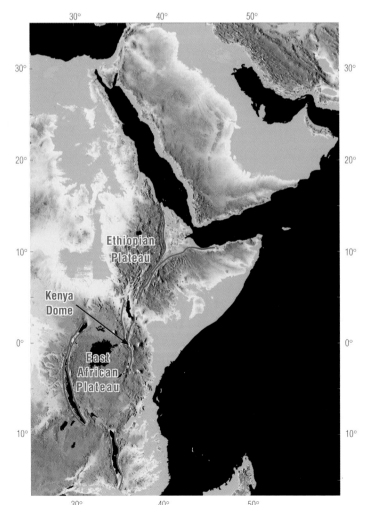

**Figure 3.2**   A larger-scale DEM of the East African Rift system (red lines) and the Red Sea. Colour coding as in Figure 3.1.

The rift is located on two major plateaux: the Ethiopian Plateau and the East African Plateau; there is a clear association between topographic elevation and the development of the rift. Indeed most of Africa's highest peaks are located close to the rift and while many of these, such as Kilimanjaro, Mount Kenya, Mount Elgon and Ras Dashan in Ethiopia, have a volcanic origin, others such as the Ruwenzori in western Uganda are non-volcanic.

●   Is the rift a continuous feature?

●   No, it is divided into three major sectors: the Kenya Rift, the Ethiopian Rift and the Western Rift.

In Ethiopia, the rift simply cuts across the Ethiopian plateau from north to south whereas across the East African Plateau the rift divides into two separate arms, the Kenya and Western Rifts. Later on we investigate why this should be.

The rift itself is defined by a 50- to 100-km-wide valley with steep sides and striking contrasts in elevation. Locally, the narrow but very deep rift valleys form some of Africa's largest lakes, such as Lakes Tanganyika, Malawi and Turkana. In other areas, the lakes have dried up to leave thick sedimentary deposits which contain much evidence of environmental change and faunal evolution. However, in Kenya and in the Ethiopian Rift, the rift fill is dominated by volcanic rocks.

The focus of this Section is the Kenya Rift, situated on the eastern half of the East African plateau (Figure 3.2). The Kenya Rift is defined as that part of the East African Rift that stretches from Turkana in northern Kenya southwards throughout Kenya and into northern Tanzania where it dies out. In more detail, Figure 3.2 shows that in addition to cutting across the East African Plateau the Kenya Rift also cuts across a smaller-scale feature known as the Kenya Dome, further emphasizing the association between topographic elevation and the development of the rift. However, the three highest points, Mount Kenya, Elgon and Kilimanjaro, are located away from the main part of the rift valley.

### 3.2.1   Summary of Section 3.2

* African topography is dominated by broad swells (plateaux) and plains.

* There is a lack of linear mountain belts on the African continent.

* The East African Rift is located on two plateaux, the Ethiopian Plateau and East African Plateau.

* The Kenya Rift cuts through the Kenya Dome, an area of unusually high elevation superimposed on the East African Plateau.

## 3.3   The geology of the Kenya Rift

Figure 3.3 summarizes the geology of Kenya which can be divided into three major units:

* the Precambrian basement;

* the Phanerozoic sedimentary cover;

* Tertiary and Recent volcanic rocks and sediments.

We deal with the geological history of each of these units in turn, emphasizing the important aspects of each to understanding the evolution of the rift.

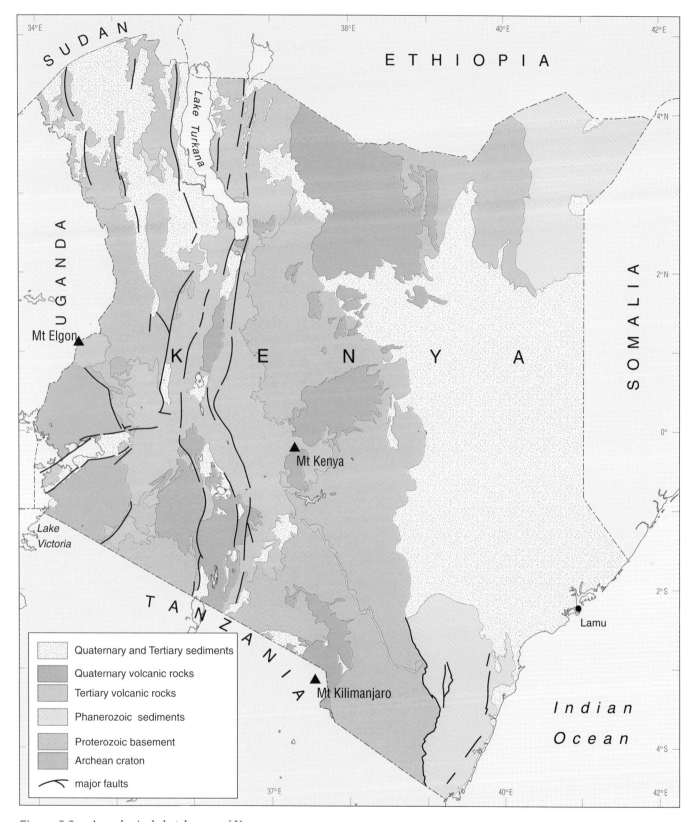

**Figure 3.3** A geological sketch map of Kenya.

### 3.3.1    The Precambrian basement

The basement to the Kenya Rift, as with much of continental Africa, comprises igneous, sedimentary and metamorphic rocks of Precambrian age. These rocks have been deformed by previous tectonic events and are generally barren of fossils, so reconstructing the geological history of the African basement has depended on radiometric dating. The picture that emerges is one of ancient continental nuclei, known as **cratons**, surrounded by younger rocks in what are known as **mobile belts** (Figure 3.4). By definition, the cratons are dominated by rocks of Archean age, i.e. older than 2.6 Ga, whereas the mobile belts are younger and either Proterozoic (2.6–0.5 Ga) or Phanerozoic (0.5 Ga to the present). Also shown are the locations of Tertiary rifts and where volcanism has occurred during the Cenozoic.

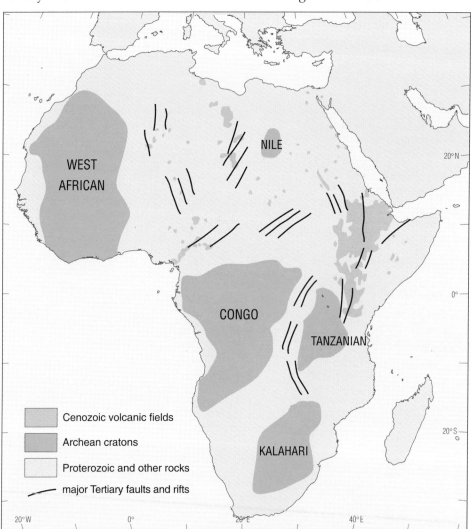

**Figure 3.4**    A sketch map of the geological basement of continental Africa, showing areas underlain by Archean cratons (>2.6 Ga) and areas underlain by so-called mobile belts (<2.6 Ga).

**Question 3.1**  Study Figure 3.4 and decide whether the rift systems and/or volcanic areas show any preference for cratonic or non-cratonic areas.

On a continental scale, the location of rifts and magmatism appears to be greatly influenced by the location of cratons. Furthermore, while some rifts have no associated volcanism, volcanic activity is almost always associated with extension.

**Question 3.2**  Do you think the association between volcanic activity and rifting has a causative link or is simply coincidence? Why?

The reason that both Tertiary rifts and volcanic rocks are concentrated in the non-cratonic areas probably lies in the differences between older cratonic and younger lithosphere. These differences were described in Block 1.

**Question 3.3** List the differences between lithosphere beneath Archean cratons and beneath Proterozoic continental lithosphere. What effect will these differences have on melting in the mantle and on extension?

Figure 3.5 shows the basement geology of the Kenya Rift in more detail, in particular the boundary between the late Archean Tanzanian craton and the much younger, Proterozoic rocks of the Mozambique mobile belt. The basement geology can be divided into three parallel bands which strike roughly NW–SE. The oldest of these in the south and west of the map is the Tanzanian craton. This

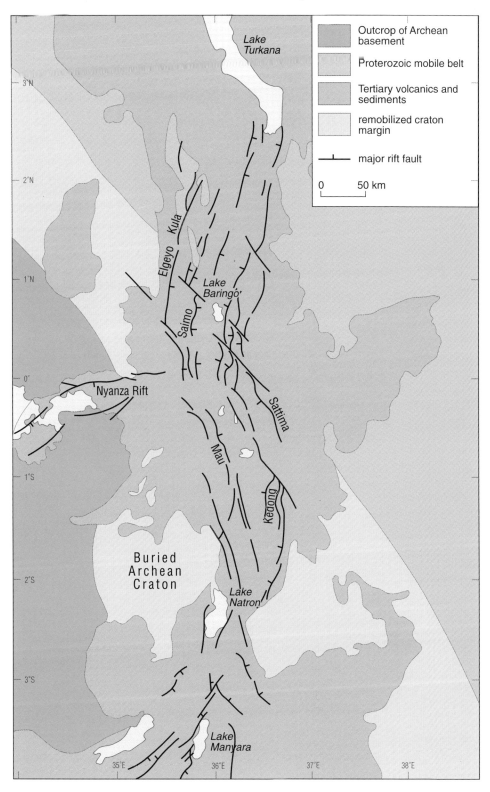

**Figure 3.5** Sketch map of the geology of the basement of the Kenya Rift. Named faults refer to major escarpments in the flanks of the rift.

contains volcanic and sedimentary rocks that range in age back to ~2.6 Ga. To the north and east is the Proterozoic mobile belt which is characterized by metamorphic and plutonic igneous rocks. These have much younger ages of between 400 and 800 Ma. In between is a zone about 100 km wide that stretches the length of the craton margin. Within this zone, rocks of the mobile belt are thrust over the margin of the craton and both cratonic and mobile belt lithologies can be found. The thrusts and shear zones in this region have been dated at 550–650 Ma and mark the site of an old continental collision zone formed during the construction of a supercontinent in the late Proterozoic. Continent collision is the subject of a later Block in this Course and we shall not investigate the details of this structure or its evolution further. It is enough now to recognize the heterogeneity of the basement to the Kenya Rift and the general NW–SE trends of the structures within it.

### 3.3.2  Phanerozoic sedimentary rocks

The Phanerozoic sedimentary rocks of East Africa are generally confined to the east coast region. Whether they were once present in central Kenya is a matter for debate but it is not significant for the evolution of the Kenya Rift. One aspect, however, is significant, and that concerns the distribution of Cretaceous rocks in northern Kenya. Quaternary sediments formed in the harsh desert conditions that exist in this region today largely obscure the older geology of northern Kenya, but geophysical exploration for hydrocarbon deposits has revealed a thick sequence of sediments lying within a deep basin, known as the Anza Rift. This structure runs from the Lamu embayment on the coast, north-west across northern Kenya and into southern Sudan. It was generated by extension in early Cretaceous times, possibly co-incident with the rifting and drifting of Madagascar away from continental Africa. The details of this event are, once again, not important but you should note the location and strike of the Anza graben as it is of some significance later on.

### 3.3.3  The Tertiary–Recent volcanic rocks

The patterns of volcanic evolution and the variations in the compositions of the different volcanic rocks types associated with the Kenya Rift are complex and the following summary and description is intended to bring out the important aspects of the magmatic evolution of the rift, possibly at the expense of some of the detail. The following Sections describe the compositions of the volcanic rocks, their ages and their distribution both in space and time.

#### Composition of volcanic rocks from the Kenya Rift

The Kenya Rift, like many continental igneous provinces, includes rocks of diverse composition. At one extreme, there are carbonatites which are virtually silicate-free, and crystallize from molten carbonate magmas rather than molten silicate magmas. The most dramatic example of this is Oldoinyo Lengai which is an active volcano in the central valley of the rift in northern Tanzania. At the other extreme are alkali-rich rhyolites with >75% silica. Figure 3.6 is a plot of total alkalis against silica for a representative suite of volcanic rocks from the Kenya Rift.

- ● Are the rocks randomly scattered across this diagram?

- ● No; they are concentrated in two diffuse fields, one at low silica (<52%) and one at higher silica (58%) and high alkali content.

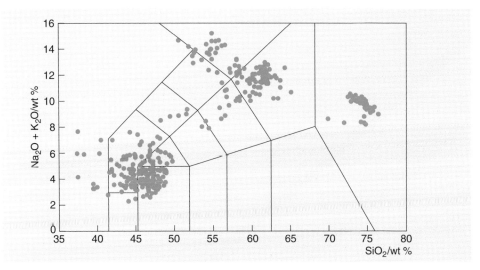

**Figure 3.6**   A plot of total alkalis versus silica content for a range of volcanic rocks from the Kenya Rift.

● What are the dominant mafic and felsic rocks of the Kenya Rift?

● The mafic rocks are dominated by basalts, trachybasalts and basanites whereas the evolved rocks are mainly trachytes with phonolites and rhyolites.

Between the compositional extremes of carbonatite and rhyolite is a large spectrum of rock compositions but these are generally of two types — mafic or felsic — and we shall be investigating the differences between these two types and their possible origins in a later Section.

● What is the term for suites of igneous rocks that include mafic and felsic but few rocks of intermediate composition?

● We have seen this rock association before in Iceland where it was described as bimodal.

The total volume of igneous material that has been erupted during the evolution of the Kenya Rift has been estimated to be about $220\,000\,km^3$, of which up to 65% is basaltic and the remaining 35% more evolved rocks such as trachytes, phonolites and rhyolites.

## Ages of volcanic rocks in the Kenya Rift

Volcanic activity is prevalent throughout the length of the Kenya Rift with historic eruptions recorded from the Barrier, south of Lake Turkana in the north, to Oldoinyo Lengai in Tanzania in the south. Each area has a complex and protracted history of volcanism and some areas reveal cycles of activity in which mafic volcanism precedes voluminous outpourings of felsic volcanics.

Figure 3.7 summarizes the details of this chronology from which two important patterns emerge:

**Figure 3.7**   A summary diagram showing the temporal evolution of magmatism in the northern and southern parts of the Kenya Rift.

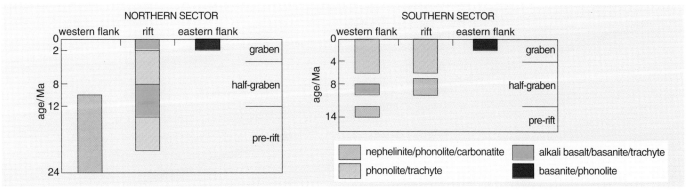

- In both the northern and the southern sectors of the rift, the earliest volcanism erupts in the west and migrates eastwards.

- Magmatism started earlier in the north (25–30 Ma) than in the south (14 Ma).

The general southward migration of magmatism with time is further supported by the observation that in northern Tanzania, magmatism did not start until 5–8 Ma. However, in all sectors of the rift, magmatism continues to the present day.

There are further patterns in the evolution of magmatism and tectonic activity in the rift and these are best illustrated using a series of maps that show the distribution of volcanic activity in relation to fault and rift basin development through time. Prior to 24 Ma, the only record of tectonic activity is the development of the Anza Rift of Mesozoic age, east of Lake Turkana and the development of extensional faults, west of Turkana, of late Paleogene age (33–24 Ma). These structures are associated with volcanic rocks that range in age from 30–25 Ma. The basins are located on thin crust of the Proterozoic mobile belt which has a generally N–S structural grain. It has been suggested that the pre-existing structures in the basement controlled the development and overall geometry of these structures.

During the early Neogene (24–11 Ma) (Figure 3.8a), we see the first outpourings of basalts and phonolites over the Kenya Dome. In addition, there was basalt, basanite and phonolite volcanic activity across the northern sector of the rift (Turkana), while to the west, over the craton and remobilized craton margin, isolated nephelinite and carbonatite centres built up. (Refer back to Figure 2.2 to remind yourself of the compositions of these different rock types.)

During the mid-Neogene (11–5.3 Ma; Figure 3.8b), volcanism decreased and rift basin morphology became more clearly defined, either by major faults or downwarps in the basement. The modern rift is beginning to take on its familiar present-day shape. By the end of this period, volcanism increased again in the form of more basalts, phonolites and trachytes but they were restricted to the newly formed rift basins.

The rift developed further during the late Neogene (5.3–1.8 Ma; Figure 3.8c) and basins developed along the whole length of the rift from Turkana in the north to Tanzania in the south. Extension was largely focused on previously active areas of faulting and downwarping. The effect of basement structure in defining the orientation of rift basins is less clear, possibly because the crust had by this time been involved in igneous activity for 20 million years and been heated and softened to such an extent that its strength was severely reduced. Throughout the rift, there was widespread volcanic activity. In the northern sector, basalts erupted from fissures producing broad shield volcanoes. By contrast, over the area of the Kenya Dome, volcanism was dominated by trachyte lavas overlain by thick volcaniclastic deposits that are thought to have erupted from now-buried calderas. In the southern sector of the rift, volcanic events were restricted to the eruption of minor nephelinites with associated trachytes, phonolites and carbonatites. There is also the first evidence of off-axis volcanism developing to the east of the rift with the growth of Mount Kenya.

Figure 3.8    Summary maps showing the geological evolution of the Kenya Rift. (a) Early Neogene, 24–11 Ma, showing the distribution of basaltic and phonolitic volcanism and the location of nephelinite–carbonatite centres and alkali basalt centres. (b) Neogene (11–5.3 Ma) showing the locations of major faults and flexures; volcanism largely restricted to rift basins. (c) Late Neogene (5.3–1.8 Ma): more intense faulting and narrowing of basins which are now more connected than during the Neogene. Faulting also extends further south into Tanzania. Development of large volume trachyte volcanic centres over the Kenya Dome, basalt and trachyte bimodal volcanism north of the dome, more alkaline centres south. First signs of volcanic activity east of the rift in Mount Kenya and the Huri Hills. (d) Quaternary (1.8–0 Ma). Development of modern rift valley and volcanism to the east of the rift (Kilimanjaro, Nyambeni, Marsabit). Rift valley is now a continuous narrow feature cutting across the Kenya Dome. Further development of highly alkaline volcanism in the southern part of the rift where it splays out into northern Tanzania.

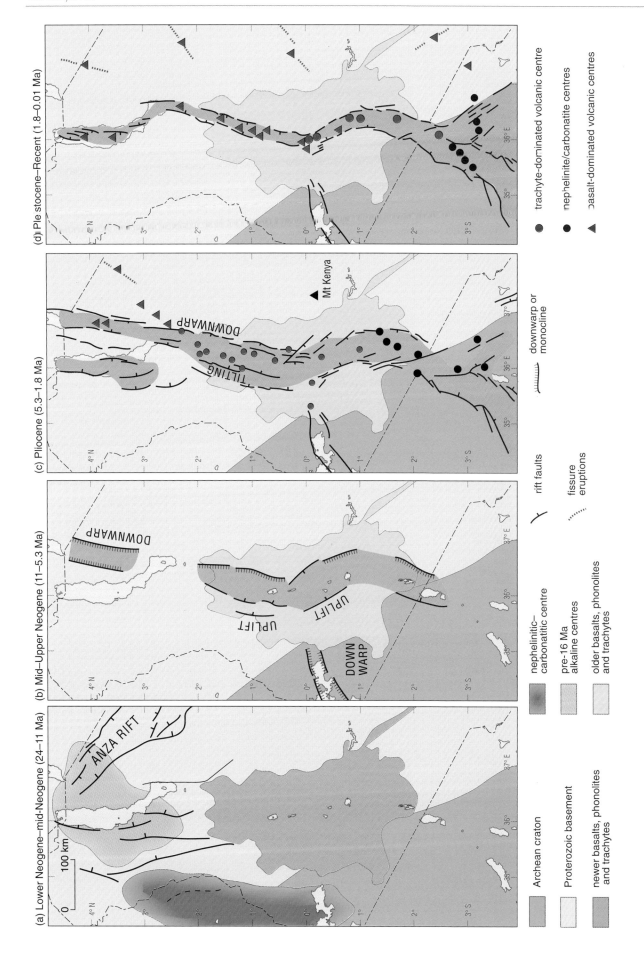

(a) Lower Neogene–mid-Neogene (24–11 Ma)

ANZA RIFT

100 km

(b) Mid–Upper Neogene (11–5.3 Ma)

DOWNWARP

UPLIFT

UPLIFT

DOWN WARP

(c) Pliocene (5.3–1.8 Ma)

DOWNWARP

TILTING

Mt Kenya

(d) Pleistocene–Recent (1.8–0.01 Ma)

Archean craton

Proterozoic basement

newer basalts, phonolites and trachytes

nephelinitic–carbonatitic centre

pre-16 Ma alkaline centres

older basalts, phonolites and trachytes

rift faults

fissure eruptions

downwarp or monocline

trachyte-dominated volcanic centre

nephelinite/carbonatite centres

basalt-dominated volcanic centres

During the Quaternary (1.8 Ma–10 Ka, Figure 3.8d), magmatism, tectonics and sedimentation became more localized in the inner trough of the Kenya Rift as we now see it. The exception to this rule, however, is in the south where the rift splays out into a broad basin, bounded by faults with NE–SW and NW–SE orientations. In the northern sector, volcanism was dominated by nephelinites and phonolites, while further south basalts and basanites were accompanied by trachyte and phonolite lavas. In the central sector, volcanism was basalt-poor and dominated by trachytes, rhyolites and phonolites, while in the south, central volcanoes are absent and volcanism is dominated by fissure-controlled eruptions. South of Lake Natron in northern Tanzania, volcanism is strongly alkaline — nephelinites associated with phonolites and carbonatites. Further off-axis volcanism to the east of the current rift continued with the development of major centres such as Mount Kilimanjaro.

> **Question 3.4** How does the original lithospheric structure affect the nature of Tertiary volcanism?

### 3.3.4    Summary of Section 3.3

* The East African Rift follows the mobile belts and seldom cuts across the craton, hence the existence of two rifts, Kenya and Western, across the East African Plateau, and only one across the Ethiopian Plateau.

* The geology of Kenya comprises three major units, the Precambrian Basement, the Mesozoic sedimentary cover and the Tertiary volcanic rocks.

* The basement can be divided into the Archean Tanzanian craton, the Proterozoic mobile belt and the remobilized craton margin lying between the two.

* The distribution of Mesozoic sedimentary rocks reflects the location of old rift basins generated during the Mesozoic, possibly co-incident with the drift of Madagascar away from the east coast of Africa.

* The Tertiary volcanic rocks of the Kenya Rift vary widely in composition but are generally mafic (basalts, basanites and nephelinites) or evolved (rhyolites, trachytes and phonolites). They show a bimodal distribution of compositions.

* Magmatism in Kenya migrates from the north to the south and from the west to the east through time.

* Nephelinites and carbonatites are largely restricted to areas of cratonic basement whereas basalts predominate on the mobile belt.

* Throughout much of its length, the rift starts as a broad extensional feature and narrows through time.

## 3.4    Geophysical studies of the Kenya Rift

Various geophysical methods inform us of the structure of the deeper layers of the crust and mantle. Those that provide the most easily interpreted data are based on observations of heat flow, gravity variations and the distribution of seismic activity. In addition, seismic refraction and reflection experiments give considerable detail of the deeper structure of the Earth and its density. The theories behind these techniques have been described in Block 1 and here you will see how they have been applied to the Kenya Rift.

Geophysical exploration of the Kenya Rift has a long history but the year 1985 marked a watershed, being the date of the first of two seismic refraction experiments that defined the large-scale and detailed structure of the Kenya

Rift. These two experiments constituted the Kenya Rift International Seismic Project, or KRISP. KRISP involved human, financial and technical resources from Germany, the UK, the USA, Kenya and many other countries, but the investment was unusually worthwhile and produced the highest resolution 3D picture of any continental rift yet obtained.

⬤ Can you think of a disadvantage of geophysical investigation over geological investigations?

⬤ Geophysics only tells us about lithospheric structure today — i.e. a snapshot in geological time. Geological investigations probe the past as well as the depths.

## 3.4.1    Geophysics and the Kenya Rift prior to 1985

The three major geophysical techniques applied to the Kenya Rift prior to 1985 involved studies of the distribution of seismicity, and of variations in gravity and heat flow.

### Gravity studies

Small but measurable variations in the Earth's gravitational field reflect variations in the density of the underlying rock layers, when corrected for topography. The resulting values for gravity can be plotted on a map and contoured to give the 'shape' of the Earth's gravity field. Figure 3.9 shows such a map of the gravity field

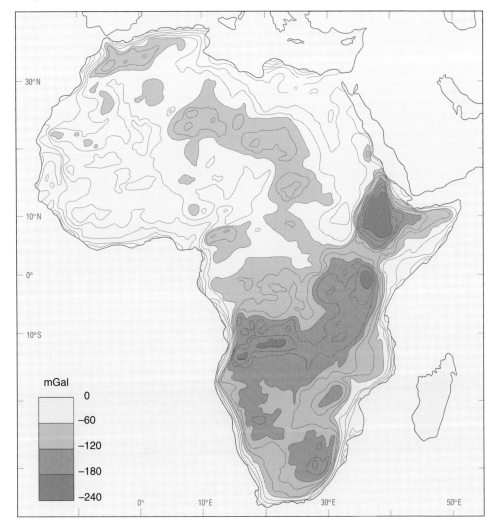

**Figure 3.9**    Bouguer gravity anomaly map of continental Africa. Contours at 20 mGal intervals.

over the whole of continental Africa. The map shows areas where gravity is greater than expected from the underlying topography (positive Bouguer anomalies) and areas where the strength of gravity is less than expected (negative Bouguer anomalies).

Question 3.5  Compare Figure 3.9 with (a) the digital elevation model of Africa (Figure 3.1) and (b) the basement map (Figure 3.4). To what extent is there a relationship between gravity and topography and basement structure?

● What does the Bouguer anomaly tell us about the density of the material underlying the East African Plateau?

● The anomaly is negative suggesting that the underlying rocks have a density which is lower than average.

The large negative gravity anomaly suggests that large regions of Africa are underlain by material of anomalously low density. So far as we can tell, there is no difference in the exposed geology between elevated and low-lying areas. From this we can only conclude that the source of the gravity anomaly must be deeper than the surface geology. In the case of East Africa, although the anomaly is most intense over the Kenya Dome, it extends across the whole East African plateau. The regional scale of this anomaly and its close correlation with topography suggests that the two have a common cause. The most likely cause of this correlation between gravity and topography is that mantle material of unusually low density underlies the elevated areas. The topography is said to be 'dynamically supported' from sub-lithospheric depths, i.e. something in the deeper mantle is pushing the lithosphere upwards, producing the East African Plateau.

The greatest anomaly associated with the East African Plateau lies across the Kenya Rift and in particular is coincident with the Kenya Dome. Figure 3.10a is a profile of the Bouguer gravity anomaly across the rift at a latitude of ~1° N. The intense negative anomaly implies that the rift is underlain by anomalously light mantle at relatively shallow depths. However, there is also a marked positive

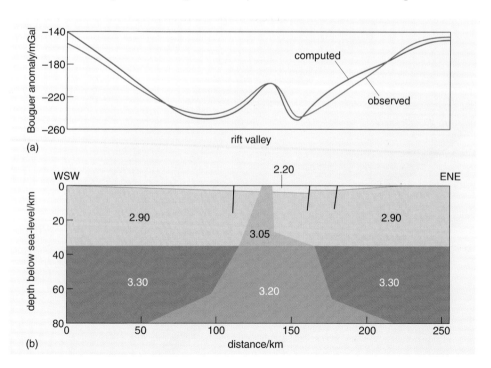

Figure 3.10  (a) Gravity profile across the East African Rift at ~1° N. (b) Interpretation of the profile in (a) based on densities of crustal rocks. Values are in $Mg\,m^{-3}$.

deviation superimposed on the regional negative anomaly, co-incident with the axis of the rift.

● If a long-wavelength negative Bouguer anomaly suggests low density material at depth, what do you think is the significance of the small positive deviation across the rift valley itself?

● The positive deviation implies an increase in density in the underlying crust while the shorter wavelength suggests that this density anomaly is smaller and shallower than the deeper density anomaly.

Figure 3.10b illustrates one interpretation of what these variations in gravity mean for the structure of the underlying crust across this whole area. The model assumes that away from the rift the crust has a typical crustal density of $2.9\,Mg\,m^{-3}$ and the mantle $3.3\,Mg\,m^{-3}$. Relative to these baseline values, there is a broad region of low density material in the mantle beneath the rift and a narrower zone of high density material at crustal levels. Similar gravity profiles across sections of the rift at other latitudes all reveal a broad negative anomaly similar to that shown in Figure 3.10. Similar axial positive anomalies, as in Figure 3.10, tend to be localized and are greatest close to central volcanoes. Where central volcanoes are absent, as in the southern part of the rift, such small anomalies are also absent.

● What does this suggest might be the nature of the dense material beneath the rift axis?

● Dense rocks or magma intruded beneath active volcanoes.

## Heat flow

Wells and boreholes sunk in and around the Kenya Rift in the search for both water and geothermal energy provide ample opportunity for heat flow measurement. We have already discussed the volcanic history of the rift and so it will probably come as no surprise to learn that heat flow is high in certain parts of the rift valley floor and especially in those areas close to active volcanoes. The average heat flow of $\sim100\,mW\,m^{-2}$ corresponds to a thermal gradient of $30\,°C\,km^{-1}$, and if this is extrapolated to 35 km depth, it implies a Moho temperature in excess of $1000\,°C$.

> **Question 3.6** What are the effects of these high temperatures on the rocks of the deeper crust and upper mantle?

On the flanks of the rift, heat flow is around half that of the rift floor at $50$–$60\,mW\,m^{-2}$ and closer to values typical of geologically inactive continental terrains. Therefore, if these thermal gradients measured at the surface and in the near-surface environment extend to depth, then the deeper layers of the crust and the upper mantle will show contrasting properties beneath the rift axis and the rift flanks.

## Seismicity

In profile and scale, the East African Rift is similar to rifts on the summits of mid-ocean ridges. Therefore it might be expected that the rift is seismically active. Figure 3.11 shows a map of seismic activity throughout a large part of the African rift system.

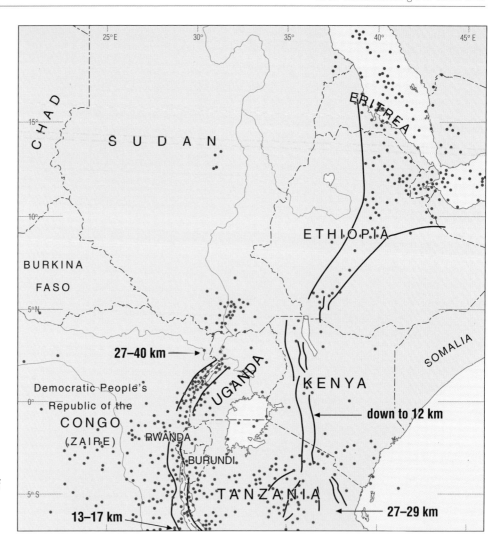

**Figure 3.11** Map of the distribution of seismic activity in East Africa (red dots), with focal depths (in km) for different regions of the rift indicated.

**Question 3.7** Describe the distribution of historic seismicity in East Africa. Does this distribution reflect other aspects of the geology of the rift?

Once again, the anisotropy of the lithosphere and the uneven distribution of magmatic activity are reflected in the distribution of seismic activity while the thermal regime controls the depth to the brittle–ductile transition within the crust.

**Question 3.8** Summarize the information on the structure of the Kenya Rift derived from geophysical investigations prior to 1985.

## 3.4.2   Geophysical investigations after 1985 — the KRISP experiments

The status of the Kenya Rift as a classic example of a magmatically active continental rift warranted further investigation of its deeper structure, and this investigation took the form of two experiments, an exploratory investigation in 1985 (KRISP85 — Kenya Rift International Seismic Project) and a more detailed experiment in 1990 (KRISP90). These two investigations deployed a network of seismic stations along and across the rift. They used seismic refraction and reflection techniques and seismic tomography to probe the structure of the crust and uppermost mantle beneath the rift. The results provided the best 3D image of the deeper structure of a continental rift so far.

## Seismic refraction results.

The KRISP85 and 90 experiments were designed to reveal crustal structure across and along the rift, as well as the structure of the rift flanks. The way this was achieved was to set up a number of recording lines with a large number of seismic recording stations at regular intervals along them (Figure 3.12) and then to set off explosions at different positions along the line to send shock waves through the crust and into the uppermost mantle. The raw data from these experiments are complex and require sophisticated computer software to make sense of them. The following Section is based on the conclusions of a number of groups, working independently of each other and using different computational techniques. However, the similarity in the conclusions from the different studies give geologists confidence that they provide a good representation of the deeper structure of the Kenya Rift.

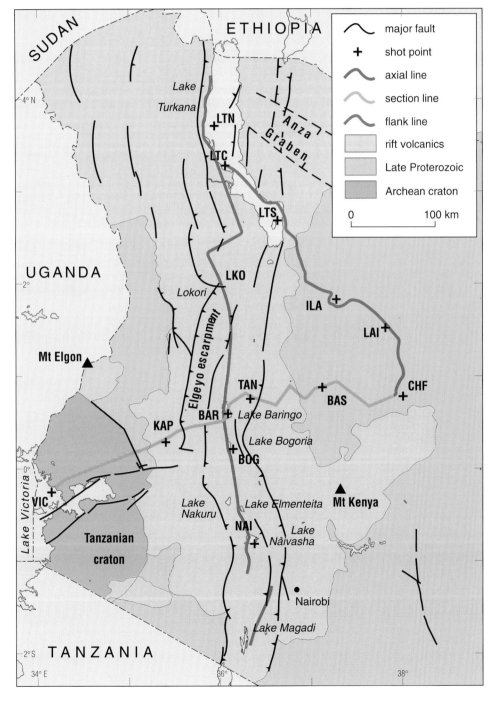

**Figure 3.12** Map showing seismic shot points and recording lines for the KRISP90 experiments. Crosses marked BAR, ILA etc. indicate the locations of seismic shot points.

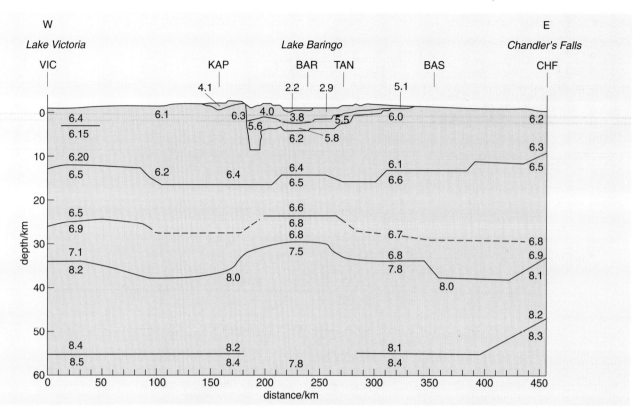

**Figure 3.13** Variation in seismic P-wave velocity with depth across the Kenya Rift at the latitude of Lake Baringo. Velocities are given in km s$^{-1}$.

Figure 3.13 shows how seismic velocities vary with depth in a vertical section across the Kenya Rift at the latitude of Lake Baringo. You should now do Activity 3.1 which will help you to relate these variations to the structure of the Kenya Rift.

## Activity 3.1  The deep structure of the Kenya Rift

Activity 3.1 leads you through the interpretation of Figure 3.13 in terms of lithologies at depth and the amount of extension across the rift. You should allow about 30 minutes for this Activity.

The Baringo seismic profile reveals an enormous amount of information about the structure of the rift beneath this location. However, do the features shown in Figure 3.13 remain constant along the rift axis? The axial shot line (red line in Figure 3.12) provides the answer (Figure 3.14). The striking feature of this

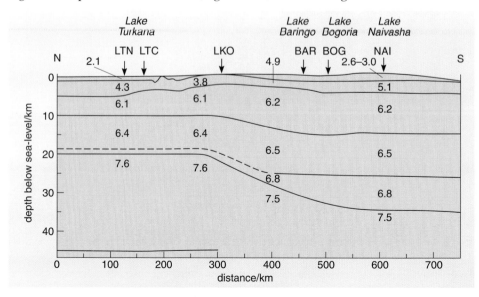

**Figure 3.14** Model of seismic velocity changes with depth in a north–south profile along the axis of the Kenya Rift. Velocities are given in km s$^{-1}$.

section is the variation in the thickness of the crustal layers along the rift. North of the cross-section in Figure 3.13 (green line on Figure 3.12, Baringo), not only does the whole crust thin dramatically from 30 km to 20 km, so too do the upper, middle and lower crust. In particular, the lower crust almost disappears. South of Baringo, the crust and its constituent layers all thicken, with the lower crust here attaining its maximum thickness of ~10 km. Clearly, the amount of crustal thinning associated with the rift appears to increase dramatically from the Kenya Dome north to Lake Turkana.

In order to test whether this extra thinning reflects extension associated with modern rifting or with an earlier episode of rifting, we can compare crustal thicknesses beneath the rift axis with those beneath the rift flanks for a number of locations along the length of the rift. This was the purpose of the KRISP90 flank line (blue line in Figure 3.12), that runs from Turkana in the north to a locality, Archer's Post, well to the east of the present rift, in the south (denoted CHF in Figure 3.12). The crustal profile along this line is illustrated in Figure 3.15. Although the north end of this line starts in the rift (station LTC), all of the stations between LTS and CHF are located on the rift flanks. Along this shot line, the Moho increases in depth sharply from stations LTC to LTS but these are within the rift and so reflect the effects of rift extension. Between LTS and LAI, crustal thickness remains more or less constant at 28 km and then between LAI and Archer's Post (CHF) the crust increases from 28 to about 33 km. There is therefore some variation in crustal thickness along this line but it is much less than that seen along the rift axis. The amount of extension increases northwards in the rift with $\beta$-values as low as 1.1–1.2 near Naivasha in the Kenya Dome, ~1.3 in the Baringo region and up to 1.6 in the Turkana region.

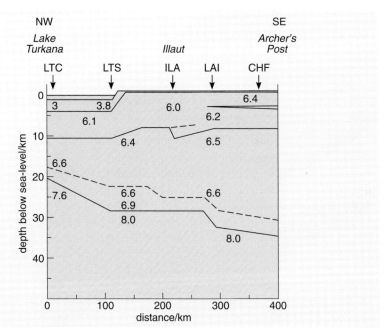

Figure 3.15 Model of seismic velocity changes beneath the eastern flank of the Kenya Rift. Velocities are given in km s$^{-1}$.

Question 3.9 How does this change in the amount of extension relate to the temporal evolution of the rift outlined in Section 3.3?

## Seismic tomography of the Kenya Rift

Seismic tomography uses variations in P-wave velocities from distant earthquakes to probe the deeper structure of the Earth. In the case of the Kenya Rift, such variations crudely chart mantle structure down to a depth greater than 150 km. Seismic refraction and reflection studies are generally concerned with two-dimensional sections through the upper layers of the Earth. Seismic

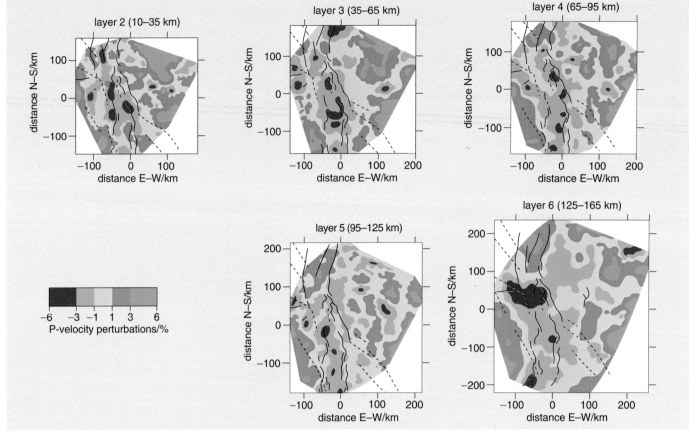

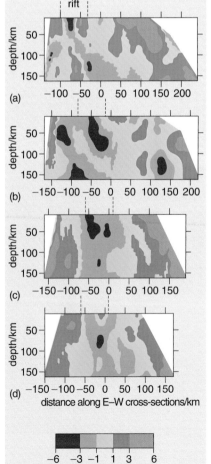

**Figure 3.16** Maps of the seismic velocity changes at six different depths beneath the KRISP90 area.

tomography is more like a CAT scan as used in medical diagnosis. It produces a series of 2D sections that can be put together to provide a 3D impression of velocity changes with depth beneath a particular region.

We have already deduced from velocity observations in the uppermost mantle that the mantle beneath the rift axis is probably hotter than that beneath the flanks. Tomography allows us to improve on that rather general conclusion and fill in yet more details. Figure 3.16 shows 2D variations in P-wave velocity in the mantle beneath the rift at Baringo for five different depths. The slices and cross-sections in Figures 3.16 and 3.17 show the percentage variations in P-wave velocity above and below a standard value, in this case 8.0 km s$^{-1}$. These vary from velocities 6% above the reference velocity to values 6% below, a total variation of 12%. All of these 'slices' show good correlation with surface geology. At all depths, P-waves travel more slowly beneath the rift than they do beneath the flanks. In layer 6, variations are much smoother and the seismically slow (i.e. hot) material present beneath the rift continues eastwards, although high velocities are still present in the western flank, underlying the Tanzanian craton.

**Question 3.10** Assuming P-wave velocity reflects temperature, what do these variations tell you about mantle structure beneath the Kenya Rift?

The boundaries between the slow mantle beneath the rift and the fast material beneath the flanks appear to be sharp and vertical. Recasting the tomographic data on a series of cross-sections of the mantle beneath the rift (Figure 3.17) shows

**Figure 3.17** Cross-sections of seismic velocity changes with depth at different locations along the Kenya Rift. The locations of the major border faults defining the rift structure at the surface are shown as vertical dashed lines.

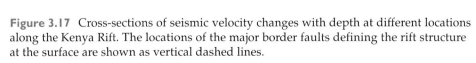

that the slowest/hottest mantle coincides with the rift axis, especially in sections c and d. The shape of this mass of slow/hot mantle is essentially a vertical sheet. The next stage in understanding the rift is to interpret these seismic velocity variations in terms of what might actually be present in the mantle.

We have already discussed the effects of density and temperature on seismic velocity but we can explore the effects of temperature in more detail using Figure 3.18a. On this diagram, variations in seismic velocity of peridotite relative to that at the temperature of its solidus ($V/V_m$) are plotted as a function of temperature. In this case, temperature is divided by $T_m$, the melting point of peridotite. Thus, values of $T/T_m > 1$ imply temperatures above the melting point, i.e. some melt is present, while at values of $T/T_m < 1$, the mantle is solid. The total velocity variation possible in solid mantle is about 6%. Seismic velocities gradually fall as temperature increases towards the melting point ($T_m$). Above $T_m$, the velocity falls very sharply because the presence of melt dramatically lowers the elastic modulus of the mantle which becomes plastic and easily deformable.

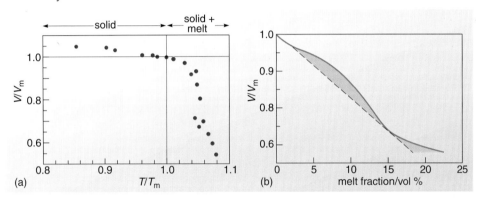

Figure 3.18 Experimental measurements of the relative changes in seismic velocity of peridotite with (a) temperature and (b) melt fraction.

● What is the total observed P-wave velocity variation beneath the rift and its flanks?

● From Figures 3.16 and 3.17, P-wave velocities can be 6% greater or 6% less than the reference value. Total variation is 12%.

● Can this variation in velocity be attributed to variations in solid mantle alone?

● No. You should recall from Activity 3.1 that there is no reason to assume that the mantle beneath the rift axis has a different density from that beneath the rift flanks. Figure 3.18a shows that temperature variations in solid mantle can produce up to 6% velocity variations. The unusually slow seismic velocities beneath the rift axis may therefore reflect the presence of partial melt.

**Question 3.11**    Figure 3.18b is a plot of relative seismic velocity, again expressed as $V/V_m$, against melt fraction. Based on this diagram and your answer to the previous question, how much partial melt is present beneath the rift axis?

Thus, seismic tomographic results place important constraints on the structure of the mantle and the amount of melt that might be present beneath the rift. Mantle melting is clearly an important part of the rifting process and we shall return to this issue in a later Section that deals with the generation and evolution of mafic magmas in the Kenya Rift.

## 3.4.3 Summary of Section 3.4

We can put all of these geophysical results together to form a composite picture of the structure of the Kenya Rift (Figure 3.19) and the important points to note are listed below:

- The whole crustal section has been thinned beneath the rift axis.

- The amount of extension decreases from $\beta = 1.6$ in the north (Turkana) to $\beta = 1.1$ in southern Kenya.

- The mantle is hot and has slow seismic velocity beneath the rift axis.

- The mantle is cold and it has fast seismic velocities beneath the rift flanks.

- The boundary between the cold and hot mantle is very steep and extends to the base of the lithosphere.

- The lithosphere is >160 km thick beneath the craton and ~125 km thick beneath the mobile belt.

- The mantle beneath the rift axis contains partial melt. The melt fraction is between 3% and 8%.

**Figure 3.19** Composite diagram showing the three sections and the seismic velocity perturbations in the mantle beneath the rift.

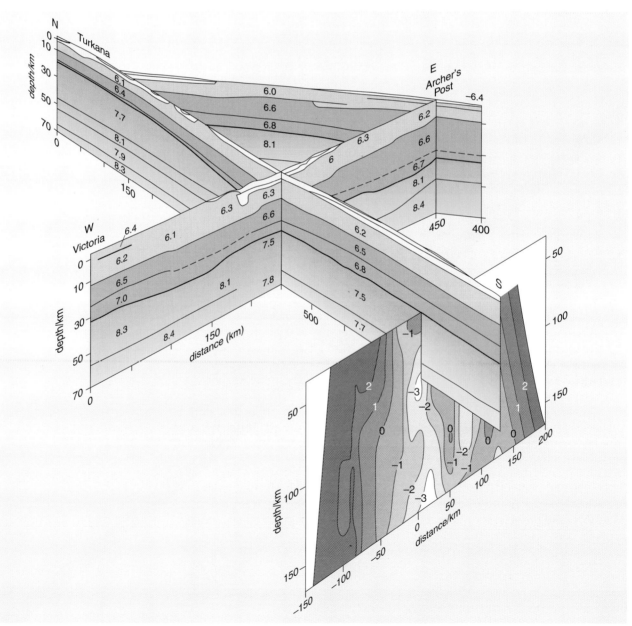

## 3.5    Extensional tectonics of the African Rift

You should recall from Block 1 and your level 2 studies that rift basins form when the lithosphere is subjected to extensional stresses. Traditionally, rifts have been thought of as symmetrical basins in which a central block subsides along two symmetrical normal faults (Figure 3.20a). Basins with this geometry are known as graben. However, structural mapping of rifts from different parts of the world shows that symmetrical basins are quite rare. In general, one side of the basin is deeper than the other and is controlled by one large fault. Such basins are described as **half-graben**, also illustrated in Figure 3.20b. A half-graben forms when two crustal blocks move along a single major fault, known as the **border fault**, to produce a wedge-shaped basin that subsequently fills with water, sediments or volcanic rocks. Deformation opposite the border fault is accommodated by flexure which, if the throw on the border fault is large enough, can develop into a series of small normal faults and **block rotation**. Note how the border fault in Figure 3.20b curves and flattens out with depth; this is known as a **listric fault**. Finally, Figure 3.20b illustrates two other terms in frequent use when describing faults: the footwall and hanging-wall.

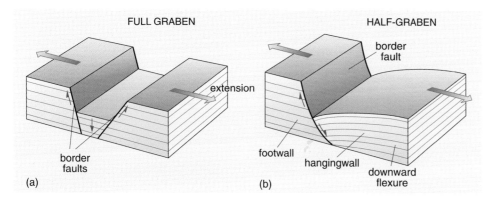

Figure 3.20  Cartoon cross-sections of (a) full and (b) half-graben structures.

What, then, is the structure of the Kenya Rift? The geological history outlined in Section 3.3 is full of expressions such as downwarps and basins but avoided the use of the terms defined in the previous paragraph. Such geological histories are based on detailed mapping and investigations of particular basins, and one of the most powerful techniques in basin analysis is seismic reflection profiling. Figure 3.21 shows the geological interpretation of a seismic reflection profile of two rift basins in northern Kenya.

- Does this profile indicate full or half-graben structures?

- The profile of the Lokichar Basin is asymmetric with a recognizable footwall, hanging-wall and border (listric) fault. It is therefore a half-graben. The Kerio Basin is also broadly asymmetric, although the eastern half has a more symmetrical, graben-like structure.

This example is from part of the northern Kenya rift where volcanic activity has been less voluminous and basins have filled with sediments. These basins may provide suitable reservoirs for hydrocarbon deposits and so oil companies are willing to invest in seismic surveys. Further south, the deep structures of rift basins have not been determined in as much detail because oil companies rarely finance such work in basins where they are unlikely to find oil. It is therefore difficult to say whether the half-graben basin is typical of the rift as a whole.

However, there have been investigations of the basins of the Western Rift. You should recall from Section 3.2 that the Western Rift has much less volcanic activity than the Kenya Rift. The individual rift basins are filled with lakes and sediments and present possible hydrocarbon reserves in an area poorly served by such deposits. Geological and geophysical investigations of the Western

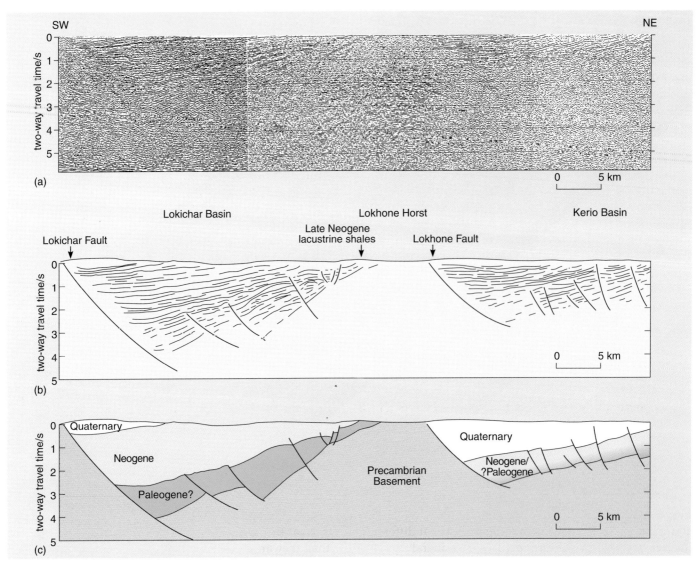

**Figure 3.21** Seismic reflection profile of the Lokichar and Kerio basins in northern Kenya. (a) Raw data from seismic experiment. (b) Interpretation of (a). (c) Structure linked in to surface outcrop.

Rift reveal that it is made up of a series of interconnected half-graben, with major border faults alternating in polarity along its length. The switching of fault polarity leads to some very complex fault geometry where the basins are connected. These complex regions between border faults are termed **accommodation zones** and Figure 3.22 shows an example from the Western Rift. Significantly, in the Western Rift, the small volume of magma that reaches the surface does so through such accommodation zones, rather than through the half-graben basins. One possible reason for this is that the large listric border faults flatten out at depth while faults in the accommodation zones could penetrate further through the lithosphere, allowing easier passage for magmas to the surface. There is certainly a greater density of faulting in accommodation zones and that may also help focus magmatism.

Whether half-graben and accommodation zones are also present in the volcanic parts of the Kenya Rift remains a matter of debate, partly because of the obscuring volume of volcanic rocks, but also because rift structure may have changed through time. Figure 3.23 illustrates one possible interpretation of the development of half-graben and accommodation zones during the middle and upper Neogene. Major scarps that alternate down the rift define the border faults, and while these faults are today seismically inactive, they are associated with thick sedimentary successions overlying the subsided hanging-wall. The current consensus is that individual basins in the Kenya Rift evolved through an initial half-graben stage.

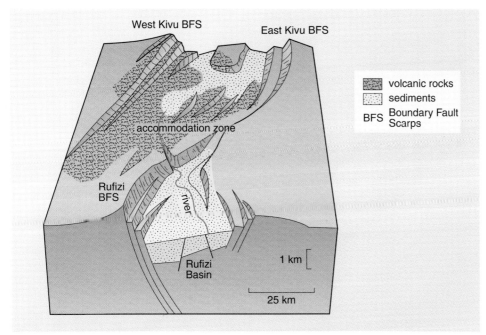

**Figure 3.22** Three-dimensional impression of an accommodation zone, based on structures mapped in the Western Rift. Outcrop of Rift volcanic rocks has been removed.

● Does the half-graben structure apply to the geometry of the modern rift?

● No: both the orientation of the major rift faults (Figure 3.8) and the deep structure of the rift defined by geophysics suggest that the modern rift is a full graben. The Kenya Rift appears to have evolved through time from a half-graben to a full graben.

The half- and full graben geometries provide an adequate description of extensional deformation in the upper, brittle crust. So, how do the lower crust and mantle lithosphere accommodate this extension? In Block 1, you read how the crust responds to stress in different ways at different depths. Near the surface, where temperatures are low, all minerals are stronger than they are at depth, where temperatures are higher. Thus, crustal materials deform by brittle failure (i.e. fracture) in the upper crust and by ductile flow at depth. The contrasting deformation styles of the shallow and deep crust have led to the development of a number of models of crustal and lithospheric extension. The following descriptions are of two 'end-member' models that successfully explain aspects of rifts and sedimentary basins in different parts of the world.

**Pure shear**. The pure shear model of extension requires that the lithosphere is attenuated uniformly along any given vertical line. To maintain such a condition, the crust fails by symmetrical faulting, block rotation and subsidence in the brittle zone and symmetrical stretching by ductile flow in the ductile zone. This model is illustrated in Figure 3.24a which shows how the whole lithosphere behaves homogeneously and is drawn out like a piece of toffee. It has been very successful in explaining the tectonic and sedimentary evolution of the North Sea.

**Simple shear**. In the simple shear model, movement occurs along a low-angle zone of deformation that penetrates the crust to the ductile layer. This zone of deformation is called a **detachment** and it forms where the crust is weakest. The model is illustrated in Figure 3.24b and can be seen to lead to an array of complex deformation structures in the upper crust. The important feature, however, is the asymmetry of the final geometry of the extended lithosphere, both in terms of the position and orientation of upper crustal faults and in the location of thinned crust and mantle lithosphere. This model developed in response to detailed mapping of strikingly asymmetric extensional structures in the Basin and Range Province of the western USA.

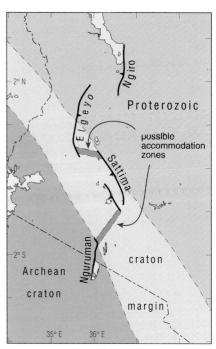

**Figure 3.23** Possible accommodation zones (dashed lines) linking major half-graben of mid to late Neogene age in the Kenya Rift. Major border faults are indicated by bold lines with tick marks on the hanging-wall. The dashed-dotted lines show national boundaries.

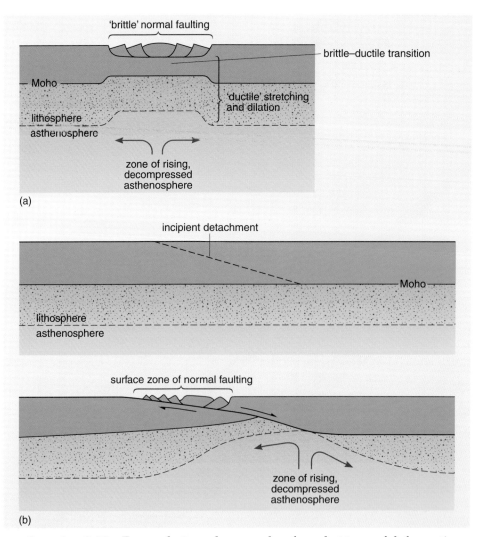

**Figure 3.24** Cartoon representations of the generation of rift valleys in the crust by (a) pure shear and (b) simple shear.

**Question 3.12** From what you have read so far, what type of deformation best describes (a) the present-day structure of the Kenya Rift and (b) the rift structure in the Neogene?

The problem remains of why the rift was initiated in the position it now occupies. You have already read how the lithosphere structure controls the locations of the Kenya and Western Rifts and how much of the Kenya Rift is located on the remobilized craton margin. This was an area of compressional deformation during the late Proterozoic that marks the site of an old continental collision zone, similar to the Himalaya–Tibet range you will read about in Block 4. Compressional deformation during the late Proterozoic generated a number of thrusts where Proterozoic crust was emplaced over the Tanzanian craton. These thrusts have low-angle geometries and represent zones of weakness through the brittle upper crust that penetrate to lower crustal depths. A popular current idea is that Tertiary extensional stresses reactivated these thrusts and partly reversed the Proterozoic tectonic movement. The reactivation of old tectonic structures is a common theme in extensional tectonics, particularly in areas where extension follows a period of compression. The unusual feature of the Kenya Rift is that compressional tectonics occurred 500 Ma ago.

Thus, the evolution of the Kenya Rift involved a long period of simple shear and the development of half-graben with major border faults possibly connected to detachments rooted in late Proterozoic thrusts in the basement. This style of extension lasted throughout the Neogene. The more recent history of the rift has been characterized by pure shear deformation and extension focused on the inner graben of the rift. Figure 3.25 illustrates one possible geometry that may

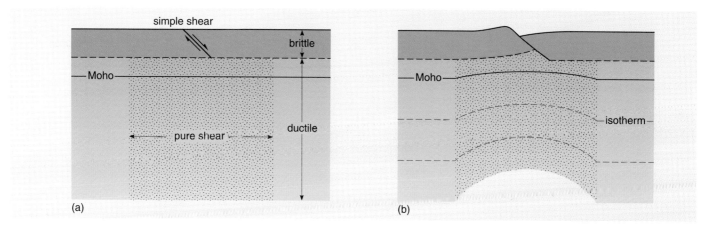

Figure 3.25 A hybrid model for the development of the Kenya Rift. (a) The initial conditions indicating brittle failure in the upper crust and ductile failure in the lower crust and mantle lithosphere. (b) The final state showing footwall uplift in the upper crust, the upward movement of isotherms in the mantle and mantle upwelling below the rift axis, leading to mantle melting below the developing half-graben.

account for this combination of characteristics. The lower crust and mantle lithosphere deform by pure shear while the upper crust extends by simple shear (Figure 3.25a). In this way, mantle upwelling occurs more or less below the rift axis, thereby allowing mantle melting through decompression below the developing half-graben (Figure 3.25b). As this system evolves, heat is advected into the crust and mantle lithosphere by migrating magma, weakening the whole lithosphere section. Further extensional stress causes the lithosphere to fail where it is weakest, which is beneath the rift basins. This brittle failure in the upper crust produces a series of antithetic faults opposite the original border fault of the half-graben, eventually generating a basin with full graben geometry.

### 3.5.1 Summary of Section 3.5

- Rifts form in response to extensional stress within the lithosphere.

- Rifts can either have a full or half-graben structure.

- The half-graben is the most common structure in extended terrain.

- The Kenya Rift has evolved through a half-graben stage into a full graben.

- Alternating half-graben are linked by complex accommodation zones which can act as pathways for magmatism.

- Lithosphere extension can occur by either pure shear or simple shear.

- A model involving extension by simple shear in the upper crust and pure shear in the lower crust and mantle lithosphere best accounts for the evolution of the Kenya Rift.

## 3.6 Petrogenesis of Kenyan volcanic rocks

In Section 2, we explored how mineralogy and composition could be used to classify igneous rocks, and how major elements can be converted to normative analyses, based on the compositions of standard minerals. In this Section, we apply these concepts to the description of volcanic rocks from the Kenya Rift and further reveal how this classification relates to the formation of magma, both in the crust and upper mantle. Igneous rocks form by melting of the mantle and crust and the subsequent evolution of those melts by processes such as fractional crystallization and mixing. As melting requires a change in the thermal regime within the lithosphere, detailed investigations of igneous rocks place further constraints on the deeper structure of the crust and upper mantle. Moreover, the compositions of igneous rocks are a direct reflection of their source regions and so geochemical studies can also tell us about the composition of the deep crust and mantle.

The volcanic rocks of the Kenya Rift are bimodal and can be broadly divided into mafic and felsic compositions. The following Sections investigate further both the mafic and felsic rocks and develop models for their generation in a continental rift environment.

### 3.6.1   Mafic rocks

Figure 3.26 shows a plot of total alkalis ($Na_2O + K_2O$) against silica ($SiO_2$) for a large number of analyses of mafic rocks from the Kenya Rift. Superimposed on this diagram is the grid defining the compositional boundaries between rock types with different names, and the broad division between alkaline and tholeiitic compositions introduced in Section 2.

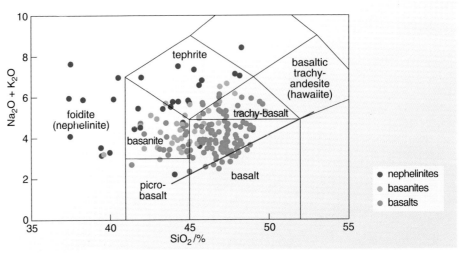

**Figure 3.26**   A plot of total alkalis against silica for mafic rocks from the Kenya Rift.

● Are the rocks of the Kenya Rift dominantly alkaline or tholeiitic?

● The majority of the analyses lie above the Hawaiian tholeiite–alkaline divide and are therefore alkaline.

● In which oceanic location are alkaline mafic rocks most common?

● Ocean islands.

Despite the Kenya Rift being an extensional environment, the dominant rock type is not tholeiitic.

> **Question 3.13**   Can you think of a reason why this might be the case?

The presence of continental lithosphere exerts a controlling effect on the compositions of erupted magmas in the Kenya Rift, but that influence is much greater than simply encouraging the generation of alkaline rather than tholeiitic rocks, as we shall see later.

The controlling influences of mantle potential temperature and lithospheric thickness on the production of basalt in an extensional environment have already been explored in Section 2. In essence, a high mantle (potential) temperature and a thin lithosphere produces maximum melt. Extension thins the lithosphere and the underlying mantle rises adiabatically into a lower pressure environment where it melts. In the case of mid-ocean ridges the amount of extension is high (the stretching factor, $\beta$, is effectively infinite) and a melt layer up to 7 km thick (the ocean crust) is produced.

In the case of the Kenya Rift, the amount of stretching is much less. Estimates of $\beta$ that we have made above from the seismic refraction profiles range from 1.1–1.6. Is this amount of stretching enough to produce melt from the mantle by adiabatic decompression?

Figure 3.27 summarizes the effects of potential temperature, lithosphere thickness and stretching factor on melt production. The three groups of sub-parallel curves define the melt thickness produced by the extension of lithosphere of varying thickness (70, 100 and 130 km) over asthenospheric mantle varying in potential temperature from a normal mantle value of 1280 °C to 1480 °C, typical of a mantle plume. Thus, if the mantle potential temperature is 1380 °C, the lithosphere originally 100 km thick and the $\beta$ factor 3, then an approximately 5-km-thick layer of basaltic melt will be produced from the asthenosphere.

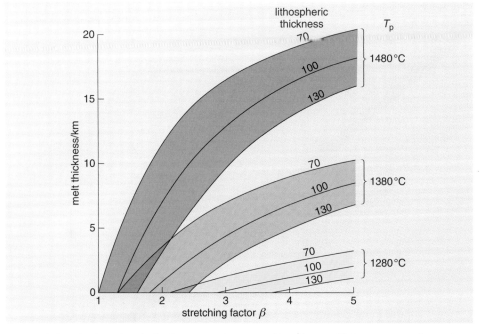

**Figure 3.27**  A plot of melt thickness against stretching factor ($\beta$) for a range of mantle potential temperatures and original lithospheric thicknesses.

● If $T_p$ is 1280 °C and the lithosphere originally 130 km thick, what is the minimum amount of extension required to induce melting in the mantle?

○ A stretching factor of 3.5–4.

Thus, for the Kenya Rift with a $\beta$ factor of <1.6 and a lithosphere that was originally 125 km thick, it is quite clear that mantle of normal potential temperature is incapable of generating any melt.

● Given these values for the Kenya Rift, what is the minimum $T_p$ required to produce melt from the mantle?

○ Only the 130 km curve for a $T_p$ of 1480 °C produces some melt if $\beta \sim 1.6$.

These preliminary observations imply the presence of anomalously hot mantle beneath the Kenya Rift for any melt to form from the asthenosphere.

Returning to the regional gravity and topography, one of the first observations we made was that the whole East African Plateau is underlain by low density material at depth and that the topography is partly maintained by upwelling in the asthenosphere. The areal extent of the plateau, and hence the dynamic support, is similar to that beneath some oceanic islands, which we know are the surface expression of mantle plumes. *Thus, both the geophysics and our preliminary findings from tectonics and magmatism suggest that the East African Plateau and hence the Kenya Rift are underlain by a mantle plume.*

## 3.6.2    The origins of the mafic margins

Moving now from the general to the specific, it is useful at this point to become a little more familiar with what might appear to be the arcane nomenclature of alkaline igneous rocks. Figure 3.26 shows us that in addition to basalts in the Kenya Rift, there are also mafic rocks with more alkali-rich and silica-poor compositions. Such rocks are called **basanites** and **nephelinites**, the latter being named after the mineral nepheline, $NaAlSiO_4$, which is an essential phase, although not necessarily the dominant one. As with alkali basalts, the sequence of crystallization in these rocks is olivine, clinopyroxene followed by plagioclase. However, as silica is reduced and the alkalis increase, the place of plagioclase is taken more and more by nepheline, which we have already encountered in the classification of basaltic rocks in Section 2.

However, nepheline has a low melting temperature, and so is amongst the last minerals to crystallize. Consequently, its presence in a volcanic rock is often difficult to detect optically because it occurs as small crystals or as a component of glass in the groundmass surrounding olivine and pyroxene phenocrysts. As we have already seen in Section 2, the potential of a lava to crystallize nepheline is revealed in the normative mineralogy which is calculated from the bulk composition, and can be used to subdivide basalts into alkalic and tholeiitic types. The division between alkali basalt, basanite and nephelinite is also defined on the basis of normative mineralogy: alkali basalts <5% Ne, basanites 5–15% Ne, nephelinites >15% Ne.

> **Question 3.14**    (a) Comment on the alkali–silica variations of the rock types in Figure 3.26 in terms of their normative characteristics. (b) Do the rock types based on normative mineralogy correspond with the field boundaries?

But why should there be such a variety of compositions? What controls the evolution of basanites and nephelinites? Are they derived from basalts or do they represent primary mantle melts in their own right? There are a number of different ways in which we can approach this but all entail an understanding of the behaviour of major and trace elements during melting and fractional crystallization.

The compositional variations from basalt through basanite to nephelinite require a drop in silica and an increase in total alkalis. Such variations are difficult to generate through fractional crystallization because olivine is the first mineral to crystallize in all of these compositions. Olivine has low silica (~40%), and so removal of olivine from a basalt with 48–50% silica will result in a fractionated liquid with even higher silica. Thus, fractional crystallization cannot be important in the generation of basanites and nephelinites.

This relationship between basalts and other alkaline mafic magmas can be investigated using the variation of compatible elements and the Mg# (magnesium number) of a rock. This term was defined in Section 2 and is analogous to the Fo content of an olivine. You should recall that when olivine crystallizes from a mafic magma it is more Fo rich than the liquid, i.e. it has a higher Mg#. Hence the liquid becomes depleted in forsterite (Mg) and enriched in fayalite (Fe), so that the Mg# of the liquid decreases. The Mg# is therefore a good **fractionation index** and can be used to investigate the variations of trace elements during fractional crystallization.

Consider two trace elements, Ni and Nb. Their partition coefficients ($K_d$-values) between olivine and melt are 10 and 0.005 respectively.

● What effect will olivine fractionation have on (a) Ni and (b) Nb abundances in a liquid?

● Ni has a high $K_d$-value and will therefore be rapidly depleted in the liquid whereas Nb has a low $K_d$-value and will be concentrated in the liquid.

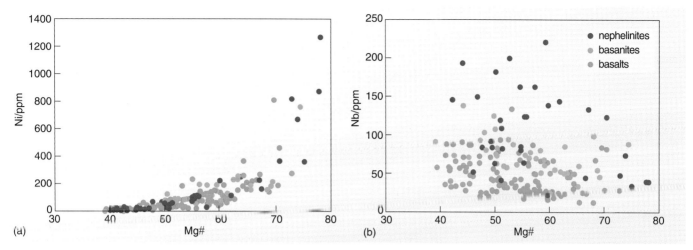

Figure 3.28   The variation of (a) Ni and (b) Nb against Mg# in mafic volcanic rocks from the Kenya Rift.

As basalts become more fractionated, Mg# falls. The $K_d$-values predict that Ni should also fall with Mg# while Nb will rise. Figure 3.28 shows the variation of Nb and Ni with Mg# for the basalts, basanites and nephelinites from the Kenya Rift. Consider first the Mg#–Ni plot; on this diagram the data for the three different rock types all lie along a broadly similar curve covering the same range of compositions, i.e. the same range of Ni and Mg# compositions. In other words there is no indication that nephelinites have lost more olivine than basalt or *vice versa*, an observation consistent with our conclusions from variations in $SiO_2$.

The variation of both major and compatible trace elements therefore indicate that alkali basalts, basanites and nephelinites are not derived from one another but represent distinct magmatic lineages. The differences between them are primary features that are characteristics of their parental magmas.

● What is a characteristic of primary mantle-derived magmas?

● You should recall from Section 2 that primary mantle-derived magmas have high Mg# (>65) and Ni concentrations (>250 ppm).

● Are the maximum Mg# and Ni concentrations of alkali basalts, basanites and nephelinites high enough for them to be primary magmas?

● Yes, there are examples of all three rock types with these characteristics.

Basalts, basanites and nephelinites are therefore derived from their own parent magmas which then begs the question as to why the mantle should produce melts of differing composition. The answer lies in the behaviour of the incompatible elements. Incompatible elements are those with very small *D*-values between mantle minerals and a basaltic melt. To deal with a specific example first, Nb is incompatible with all of the major mantle minerals (olivine, pyroxenes, spinel and garnet).

Question 3.15   Using Equation 2.1 (p.33) for partial melting, calculate the concentration of Nb in a magma derived by 1%, 5%, 10% and 25% melting of the mantle. Assume a *D*-value of 0.005 and a concentration of Nb in the mantle of 0.5 ppm.

The answer to Question 3.15 shows that the concentration of Nb varies by more than an order of magnitude between 1% melt and a 25% melt, and the variation is more pronounced in the small melt fractions. How do these calculated concentrations compare with those observed in the Kenyan samples? Although the data appear scattered in Figure 3.28b, the overall tendency is for the nephelinites to have higher contents of Nb at a given Mg# than the basanites which have higher Nb contents than the basalts.

● What are the implications for the origins of these magma types?

● The abundances of Nb suggest that the nephelinites are produced by the smallest melt fractions, the basalts by the largest melt fractions and the basanites by intermediate melt fractions.

While calculating the relative melt fractions during magma evolution is fairly straightforward, calculating the absolute (actual) melt fractions is less simple. However, the following question further illustrates the magnitude of the melt fractions that might be involved in the generation of alkali-rich basalts, basanites and nephelinites.

> **Question 3.16**   Assuming the same values for the mantle $D$-value and concentration of Nb as in Question 3.15, what melt fractions do the following Nb concentrations imply for the different magma types: nephelinites 75 ppm Nb; basanites 45 ppm Nb; basalts 25 ppm Nb?

● What other technique has given an indication of melt fraction in the mantle?

● Seismic tomography (see Section 3.4).

● How do the values calculated in Question 3.16 compare with estimates of melt fraction beneath the rift from seismic investigations?

● Seismic investigations suggested between 3% and 8% melt beneath the rift axis, whereas those calculated in Question 3.16 are smaller.

Estimates of melt fractions based on geochemistry are generally smaller than those derived from geophysical investigations, perhaps by an order of magnitude. Let us see if we can reconcile these two lines of evidence.

● What major assumptions have been made in these melting calculations?

● The values of the partition coefficient and the source concentration of Nb.

The discrepancy between the melt fractions estimated from the KRISP data and those estimated from geochemical approaches could be due to the application of incorrect $D$-values or an incorrect estimate of the source concentration of Nb. Of these two, it is the source concentration of Nb that has the major effect on the calculated melt fraction and is the parameter that is least well known. Partition coefficients are increasingly well determined from experiment and observation but analyses of mantle materials reveal at least an order of magnitude variation in Nb and other incompatible elements in the upper mantle. The value of 0.5 ppm Nb is an average of many analyses of mantle samples and so our estimate of melt fraction could be a factor of 5 too small. If this is the case, then the two approaches converge on similar results, although it implies that the mantle below the rift is enriched in Nb (and perhaps other incompatible elements).

## 3.6.3   More about mantle melting

One of the most convenient ways to display trace element data for a basaltic rock is to divide each element concentration by its concentration in the mantle and then to plot the normalized abundances of a range of trace elements in their sequence of incompatibility in the mantle (Figure 2.18). A glance at such a plot for three mafic magmas from the Kenya Rift (Figure 3.29) reveals important differences between them. The most obvious feature of this diagram is that all of the profiles are sub-parallel — those with high Nb abundances also have higher concentrations of the other incompatible elements. This reinforces

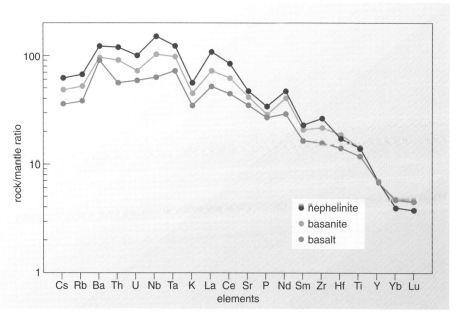

**Figure 3.29** Average trace element concentrations in Kenya Rift mafic volcanics, normalized to abundances in the primitive mantle.

the conclusions arrived at from a consideration of Nb alone that nephelinites are derived from smaller melt fractions than basanites, and basanites from smaller melt fractions than basalts. The second feature is that the shapes of the profiles are remarkably similar to those of ocean island basalts and related magmas. We know from Section 2 that these are derived from mantle plumes. This lends further support to our hypothesis that a mantle plume lies beneath the Kenya Rift and East African Plateau.

There are, however, some more subtle variations in trace element abundances. The first of these concerns the more compatible elements to the right-hand side of the diagram, Y and Yb. Rather than varying with Nb, these elements tend to have the same concentration range in each of the magma types.

⬤ Which other element behaves like this?

⬤ Ni.

Ni is a compatible element and the more limited variation of Y and Yb compared with Nb implies that these two elements are behaving more compatibly than Nb.

> **Question 3.17** Study the list of partition coefficients for various trace elements between different mantle minerals and mafic liquids in Table 2.7. With which mineral(s) are Y and Yb compatible and at what pressure is this mineral(s) stable?

This observation tells us that melting must have extended to these depths, particularly for those magmas with high Nb and other incompatible element concentrations but low Y and Yb, in other words those compositions with high Nb/Y and Nb/Yb ratios. Once again, we can tie this observation in with variations in major elements, in particular silica. We know from Section 2 that silica varies with depth of melt generation and that low-pressure melts have high silica and high-pressure melts low silica. Therefore, if our conjecture about the role of garnet is correct, then those rocks with low Nb/Y should have higher silica contents than those with high Nb/Y. Figure 3.30 shows the covariation of silica with Nb/Y in the Kenyan mafic rocks. Certainly, the basalts and basanites show a general negative trend and whereas the nephelinites are more scattered, they too have higher Nb/Y and lower silica.

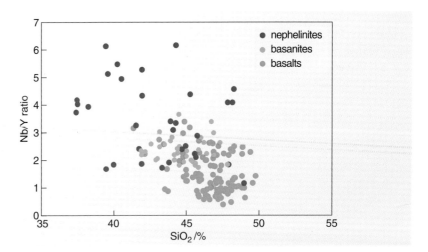

**Figure 3.30** Covariation of silica with the Nb/Y ratio in mafic volcanic rocks from the Kenya Rift.

The average silica contents of each magma group are: nephelinites 42.7%, basanites 44.7% and basalts 46.7%.

> **Question 3.18**   Using Figure 2.17, what pressures of melt generation do the average silica contents of the basalts, basanites and nephelinites imply?

Finally, we can use these depth estimates in combination with conclusions regarding melt fraction to tell us something about the variation of melt fraction with depth. We know from incompatible elements that the relative melt fraction increases from nephelinite through basanite to basalt. Therefore, the amount of melt generated in the mantle beneath the rift decreases with increasing depth. Once again, this is consistent with models of decompression melting in which hot material rises from a high pressure to a lower pressure regime (Figure 2.12). However, the lack of tholeiites in the rift suggests that melts are not produced at depths of less than about 40 km. Although the average basalt has 47% silica, primary basalts do occur with up to 49% $SiO_2$, equivalent to a depth of 40 km. Thus, the top of the mantle melting zone may be as shallow as 40 km.

● How does this minimum depth compare with the crustal thickness beneath the rift axis?

○ If you study Figure 3.13, the base of the crust below the rift axis is at a depth of 30 km. Therefore, the calculated pressure of melting results from the overburden of the crust plus a thin remnant of mantle lithosphere.

The presence of melts derived from such shallow depths indicates that hot mantle has penetrated to depths much shallower than would be expected from the effects of extension alone. Remember, the maximum $\beta$ factor along the rift is 1.6. Assuming an original lithospheric thickness of 125 km and homogeneous thinning, then the stretched lithosphere should still be 125/1.6 km or about 80 km thick. Therefore, while the average basalt composition suggests melting from a depth of 60 km, broadly consistent with the amount of extension across the rift, the total variation in basalt composition implies that melting occurs at depths almost as shallow as the Moho.

This picture may seem confusing and indeed it is, because even at the time of writing there is no clear consensus on the actual minimum depth of melting in the mantle beneath the Kenya Rift or the details of how the melt fraction varies with depth. What is clear is that melts are produced over a range of pressures, that the basalts are derived from the shallowest depths, the basanites from intermediate depths and the nephelinites from the greatest depths, and that the melt fraction decreases from a maximum at 40–60 km and dies away to nothing at 100–150 km. It should also be remembered that while we have described the generation of mafic magmas as if there were three discrete compositions, the data, as in Figure 3.30, imply a continuum of magma compositions between the most silica-rich basalt and the most silica-deficient nephelinite. These magmas are produced by a combination of variable melt fractions and mixing between melts as they are generated.

> **Question 3.19** Is this description of the melting regime consistent with decompression melting in a mantle plume? Give reasons for your answer.

### 3.6.4 Generation of felsic magmas in the Kenya Rift

In addition to producing large volumes of basalt, volcanism in the Kenya Rift has also produced significant amounts of felsic magmas, ranging in composition from nepheline-normative phonolite, both nepheline and quartz-normative trachytes and alkali-rich rhyolites (also known as comendites and pantellerites). The petrology of this volcanic activity is described in the following Activity that covers this aspect of the magmatic evolution of the Kenya Rift.

## Activity 3.2   Felsic magmas in the Kenya Rift

Activity 3.2 is video-based and describes the field relations of felsic volcanic rocks from different localities in the Kenya Rift and explains how they may have evolved. You should allow about 45 minutes for this Activity.

The presence of large magma bodies high in the crust and partial melting zones in the deep crust within the rift are the product of, and amplify, high heat flow from the underlying mantle. Heating the crust to such an extent that it melts also weakens it. If further extensional stresses are applied, then the crust will fail where it is weakest, which is across the rift. More extension produces more melt in the mantle which advects more heat into the crust and so on. This inter-relation between extension, magmatism and crustal/lithospheric strength produces a weak zone within the lithosphere, focusing deformation on a narrower and narrower zone. This is why the rift, initially a broad feature up to 150 or even 200 km across, has narrowed so that the active rift is now in places less than 50 km across. But this analysis leaves the question of the source of the extensional stress unanswered which is the subject of the next Section.

### 3.6.5 Summary of Section 3.6

- The volcanic rocks of the Kenya Rift show a bimodal distribution with large volumes of mafic rocks, large volumes of felsic rocks but few intermediate compositions.

- Mafic rocks are generally alkaline and silica undersaturated (Ne-normative), and range in compositions from alkali basalts, basanites and nephelinites.

- Alkali basalts, basanites and nephelinites are not related to each other by fractional crystallization but represent distinct mantle-derived magmas.

- Alkali basalts, basanites and nephelinites are derived from the mantle as a result of variable degrees of melting at different depths; nephelinites from the deepest and smallest melt fractions, basanites from intermediate depths and melt fraction and alkali basalts from the shallowest depths and largest melt fractions.

- Variation in composition within the different mafic rock suites is dominated by olivine fractionation.

- Felsic rocks range in composition from rhyolites, trachytes to phonolites.

- The felsic rocks of the Kenya Rift can be derived both by partial melting of a suitable source and by fractional crystallization of a suitable parental basalt.

## 3.7   The geodynamics of the African Rift

We have already briefly mentioned that the African continent is largely devoid of linear mountain belts because the African Plate lacks destructive plate margins. Figure 3.31 shows the African Plate, surrounded to the east, south and west by mid-ocean ridges. In north-east Africa, the plate boundary is marked by a constructive margin in the Gulf of Aden and by the Red Sea between Africa and Arabia. The latter is also a constructive margin and we shall be exploring its evolution in the next Section. These two features are linked to the East African Rift in the Afar Depression in north-east Ethiopia.

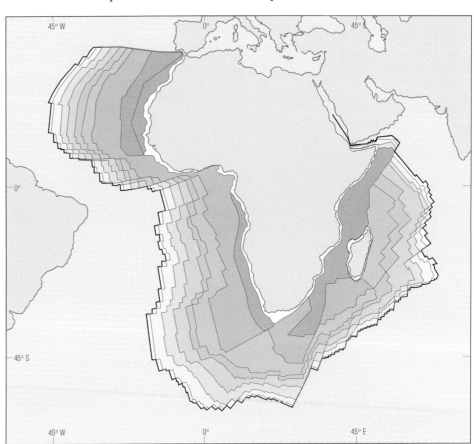

**Figure 3.31**   A map of the African Plate showing the age structure of the oceanic crust. Darker shades indicate progressively older crust and each band represents 20 million years.

- If the East African Rift is linked to the mid-ocean ridge system in the Afar, what would this feature be called?

- A triple plate junction, type RRR.

This possible link between the East African Rift and the mid-ocean ridge system has led to suggestions that perhaps Africa is breaking up along the line of the rift which will one day evolve into a new 'African' Ocean. While such a future may or may not await the African continent, it is pertinent now to investigate the nature of rifts, how they form and the controls on their structure.

The constructive plate boundaries that almost surround the African Plate are characterized by extensional features. This implies that the African Plate should be under a state of near radial compression. Only in the north might there be any extensional stress applied to the plate through continental collision and subduction of the African Plate, as happens beneath the Aeolian and the Aegean arc systems in the Mediterranean region. Constructive plate boundaries effectively isolate the plate from external stresses, so deformation within the African Plate must be driven by stresses generated entirely internally. But how can internal plate stresses be generated, especially extensional stresses large enough to overcome the effects of the mid-ocean ridges?

When plate boundary forces are removed, the dominant force acting on a plate is gravity. It is, of course, gravity that drives descending slabs, which are cold and dense relative to the warm asthenospheric mantle, that produces slab-pull, the dominant driving force for plate motions on Earth. However, without the effect of slab-pull, gravity still affects the behaviour of plates and generates stresses within them. One way of investigating the effect of gravity on a plate is to calculate the distribution of **gravitational potential energy (GPE)** across it. Such calculations are well beyond the scope of this Course but have been carried out for the African Plate, and the following paragraphs describe the results of those calculations.

In essence, the potential energy of any part of a plate depends on its elevation above or below sea-level, and the mean density of the plate at that location. Elevated regions of continents have high GPE whereas the older deeper parts of the surrounding oceanic lithosphere have low GPE. In between, low elevation continental crust and oceanic lithosphere of moderate depth have similar GPE because the higher density of the oceanic lithosphere compensates for the higher elevation of the less dense continents.

Figure 3.32 shows these differences in GPE. The bottom part of this diagram plots the elevation relative to sea-level of a plate from an ocean ridge, through an ocean basin across the continental margin and into the continental interior. This represents the topography from W to E of the African Plate from the Mid-Atlantic Ridge, across the Atlantic Ocean and into the interior of the African continent. The top half of the diagram shows the distribution of the GPE across the same section. Elevated regions of Africa should have high GPE and could therefore extend if the lithosphere had insufficient strength to support itself. By contrast, the deep ocean basins have the lowest GPE. The variation of GPE

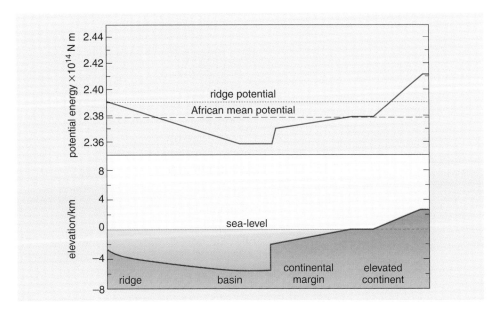

Figure 3.32 Sketch profiles of the topography and the gravitational potential energy (GPE) across a hypothetical section through the African Plate.

across the plate is shown with reference to two important values. The first is the mean GPE of the African Plate as a whole and the second is the GPE of the mid-ocean ridge system. All those regions with a GPE greater than that of the mean are in a state of tectonic tension while those with lower energy are under compression. Note that the Mid-Atlantic Ridge in this model is above the mean GPE and this contributes to the driving force for extension across it.

Topography clearly has a dramatic effect on the potential energy of different parts of a plate. To understand internal plate stresses, we need to grasp what causes variations in elevation. In the case of Africa, we now know that the greatest elevations result from the effects of mantle upwelling — the **dynamic topography** supported by mantle plumes. In the oceans, however, the picture is somewhat different. Variations in the depth of the mid-ocean ridge system are also directly related to the thermal structure of the underlying mantle; however, it is clear from the profiles in Figure 3.32 that it is not the ridge GPE that affects the mean plate GPE but the GPE of the deep ocean basins.

● What controls the depth of the ocean basins?

● Their age. As ocean lithosphere ages, it cools and subsides into the mantle.

● What would happen to the mean GPE in Figure 3.32 if the ocean basin was younger and therefore shallower than that shown?

● Because more mass is at a higher elevation, the mean GPE must increase.

In the case of the African Plate, we can see in Figure 3.31 the ages of the different areas of the ocean basins surrounding continental Africa.

● What is the age of the ocean lithosphere immediately adjacent to the east and west Africa coasts?

● For the most part, it is Lower Cretaceous or Jurassic, older than 140 Ma.

The oceanic lithosphere surrounding continental Africa is amongst the oldest of any of the ocean basins. As oceanic lithosphere spreads away from the ocean ridges, it cools, and thickens as a thermal boundary layer. The ocean floor gradually subsides because of the increase in density caused by cooling and the mean GPE of the whole plate decreases because the mean plate elevation also decreases. We therefore have two important processes that contribute to the topographic profile of the African Plate: dynamic topography from mantle plumes and thermal subsidence of oceanic lithosphere as it migrates away from the MOR system.

The mean plate GPE thus evolves with time but at present most continental regions of Africa, where altitude is above sea-level, have a GPE close to the mean value for the African Plate. However, as altitude increases so does GPE and for a mean African altitude of 650 m the GPE is above the average for the whole African Plate. Continental Africa is therefore under an extensional stress. In the most elevated regions, where the mean elevation is 2 km, the extensional stress reaches a maximum, which is higher still than that exerted by the mid-ocean ridge system.

The direction and magnitude of these extensional forces can be calculated and the results for Africa are shown in Figure 3.33. It is clear from this that the deep ocean basins are in a state of compression, whereas the mid-ocean ridges are under tension, as are the continents. The maximum predicted extensional stress occurs over the highlands of eastern Africa, coincident with the African Rift.

To determine whether or not continental lithosphere will extend when subjected to these so-called buoyancy forces requires estimates both of the strength of the continental lithosphere and the magnitude of these forces. The strength of the continental lithosphere has been estimated to be in the range of $3–4 \times 10^{12}\,\mathrm{N\,m^{-1}}$.

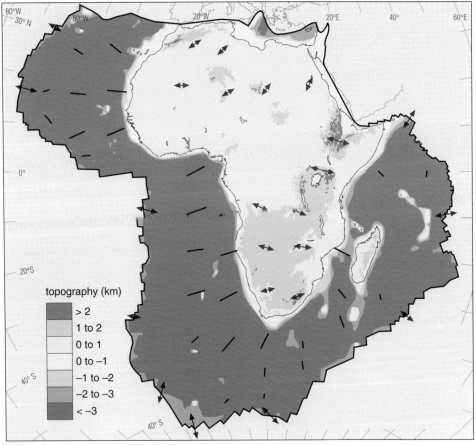

**Figure 3.33** The magnitude and direction of calculated stresses across the African Plate. Red arrows indicate estimated tensional stress and black bars compression. The lengths of the arrows and bars indicate the magnitude and their orientation the direction of the stress.

The principal forces driving plate motions are ridge-push ($3-4 \times 10^{12}\,\mathrm{N\,m^{-1}}$) and slab-pull (up to $10^{13}\,\mathrm{N\,m^{-1}}$). The latter is the larger by a factor of two to three and in the case of plates with significant lengths of actively subducting margins, it dominates the forces acting within and across the plate. However, in the case of a plate with only short lengths of active subduction, the dominant plate margin force is that exerted by the ridges, i.e. ridge-push.

- What is the difference between the GPE of the continental interior and the African mean GPE in Figure 3.32?

- Maximum continental GPE = $2.41 \times 10^{14}\,\mathrm{N\,m}$ whereas the African mean GPE = $2.38 \times 10^{14}\,\mathrm{N\,m}$. The difference is $0.03 \times 10^{14}\,\mathrm{N\,m} = 3 \times 10^{12}\,\mathrm{N\,m}$.

The increase in the mean density of the ocean plate as it ages produces an extensional force of a similar magnitude to the force exerted by mid-ocean ridges ($1-3 \times 10^{12}\,\mathrm{N\,m^{-1}}$). Additional extensional forces result from elevated continental topography, which are of a similar magnitude ($1-2 \times 10^{12}\,\mathrm{N\,m^{-1}}$) when the mean surface elevation approaches 2000 m, as it does in East Africa. Adding these two gravitational forces together produces a total extensional force of $2-5 \times 10^{12}\,\mathrm{N\,m^{-1}}$. This is comparable to that required to overcome the extensional strength of the continental lithosphere. However, over the site of a mantle plume, the lithosphere will be both heated and thinned, initially by conduction from below and later by advection of heat by the movement of magma into and within the lithosphere section. Hence, the mantle plume is important for two reasons: first in the development of dynamic topography, which contributes to GPE, and secondly in heating and weakening the lithosphere in the same place that the maximum GPE is generated. Thus, rifting along the length of the East African Rift system is the result of the combination of a number of effects related to both the presence of mantle plumes and the age structure of the African Plate itself.

Ultimately, extension across the African Rift is limited by ridge-push. So, given present plate configurations it is unlikely that an 'African Ocean' will develop. However, 30 Ma ago, plate configurations were significantly different and Section 4 focuses on the evolution of the Red Sea which originally started as a rift and evolved into a new ocean basin.

### 3.7.1   Summary of Section 3.7

- The African Plate is surrounded by constructive plate boundaries which exert a modest compressional force (ridge-push).

- Extension across the East African Rift must therefore be driven by forces generated within the African Plate.

- The African continent is surrounded by old, cold and dense oceanic lithosphere which exerts an extensional force on the ocean ridges and the continental lithosphere.

- The continent is in a general state of tensional stress.

- This extensional stress is large enough to overcome the strength of the lithosphere only in the most elevated parts of the continent.

- These elevated areas are also magmatically active and have a high geothermal gradient so that the lithosphere is weaker and thus fails more easily under tension.

- The mantle plume has a two-fold effect in that it provides dynamic elevation and thermally weakens the continental lithosphere.

## Objectives for Section 3

Now that you have completed this Section, you should be able to:

3.1   Understand the meaning of all the terms printed in **bold**.

3.2   Outline the basement geology of Africa and how that affects the location of Tertiary rifts and volcanism.

3.3   Summarize the geology of Kenya and the evolution of the Kenya Rift.

3.4   Using relevant geophysical information, describe the deeper structure of the Kenya Rift.

3.5   Discuss the differences between the pure shear and simple shear models of crustal extension in the upper and lower crust.

3.6   Give an account of the petrogenesis of mafic rocks in the Kenya Rift based on major and trace elements.

3.7   Describe the plate tectonic setting of the Kenya Rift and discuss the sources of extensional stress across the African Plate.

3.8   Summarize the evidence for the presence of a mantle plume beneath the Kenya Rift and the East African Plateau.

# 4  The Red Sea

## 4.1  Introduction

Your study of the Kenya Rift has shown how the continental lithosphere responds to extensional forces generated by contrasts in gravitational potential energy across a single plate. However, the driving forces within a plate are self-limiting. A mantle plume can generate only a restricted amount of uplift while old ocean crust will eventually be subducted into the mantle, changing the plate configuration and the distribution of gravitational potential energy (GPE) across the plate. Intraplate stresses can produce a rift but they cannot lead to continental break-up and the development of a new ocean. Yet such events do occur and the best-documented example of recent continental break-up is the

**Figure 4.1**  Major plate boundaries and tectonic features of the African Plate.

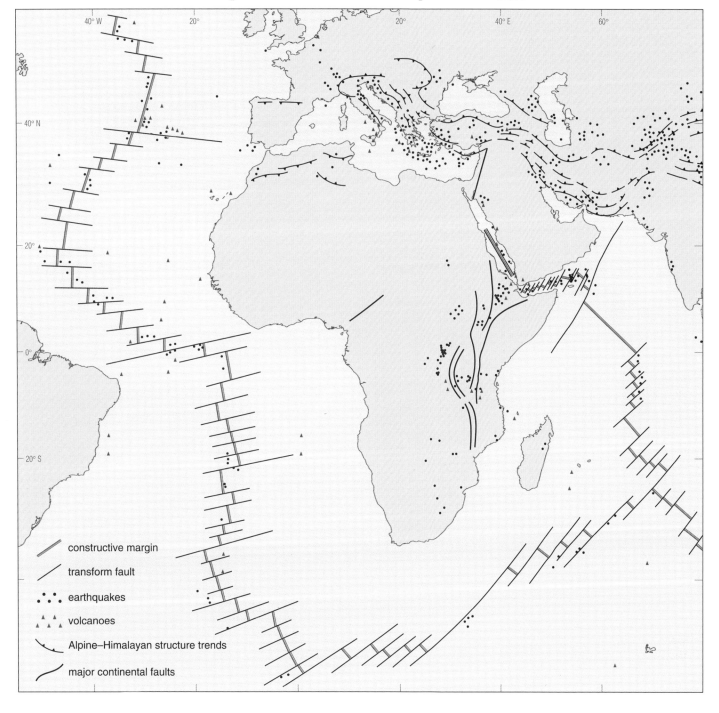

constructive margin

transform fault

earthquakes

volcanoes

Alpine–Himalayan structure trends

major continental faults

Red Sea. The following Section continues the investigation of continental break-up by describing some of the important aspects of the geological, tectonic and magmatic history of this youngest of ocean basins.

## 4.2 The regional geology and tectonics of the Red Sea

The Red Sea differs from major ocean basins in having an elongated shape with a 10 : 1 length-to-breadth ratio (Figures 3.2 and 4.2). The shape alone tells us that it is probably a young ocean basin, because the width of a spreading ocean obviously increases with age. Figure 4.1 (on the previous page) shows the position of the Red Sea in relation to plate boundaries around Africa.

> **Question 4.1** Now study Figure 4.2 (below). Examine the eastern and western margins of the Red Sea. Is there any degree of 'jigsaw' fit which could suggest that Africa and Arabia were once joined together? (Trace the outline of the Arabian coast on transparent paper and attempt to fit it against the African side.)

From your answer to Question 4.1, you should have noticed that in the middle section the fit is excellent; problems lie at the northern and southern extremities. The movement of Arabia relative to the African Plate is to the north-east with the amount of separation increasing slightly southwards. Nevertheless, at the northern end the width of the Gulf of Suez is too narrow, although this can be accounted for by considerable lateral displacement along the Gulf of Aqaba (see Figure 4.2). Also, Africa and Arabia overlap in the south if the Red Sea is closed right up. Towards the south, the Red Sea narrows until at the southern entrance it is only 35 km across. The simple coastal fit hypothesis does not work well here either. We shall return to both these problem areas later, by taking a closer look at the geological structure and evolution of the area.

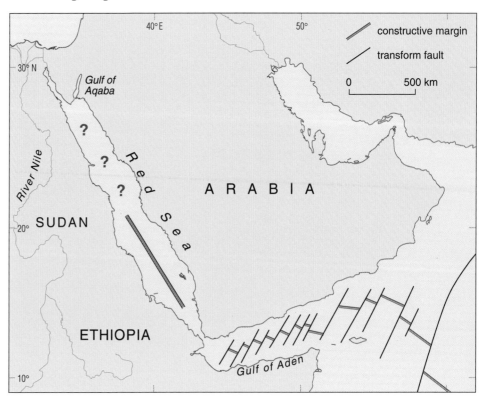

**Figure 4.2**   Red Sea area showing position of major constructive margins in the Red Sea and Gulf of Aden.

## 4.2.1    Regional tectonics

Before we focus on the Red Sea itself, we need to consider the regional
tectonics of Arabia and Africa which provide a framework for the evolution
of the Red Sea. Return to Figure 4.1 which shows the earthquake foci and
active volcanoes associated with the major tectonic features in and around
Africa.

> **Question 4.2** (a) How does the character of the plate margin change at
> the western extremity of the Gulf of Aden (located between the southern
> coast of Arabia and the Horn of Africa)? (b) What is the spatial
> relationship between the Red Sea, the Gulf of Aden and the East African
> Rift system? Sketch the plate movement vectors around the point where
> these three features intersect.

You should recall that any combination of the three types of lithospheric plate
boundaries (constructive, destructive and conservative) can come together at
a triple junction. From Question 4.2, you can see that the triple junction at the
southern end of the Red Sea is a triple rift or RRR junction. However, the East
African Rift has not developed into a marine area as have the Red Sea and
Gulf of Aden. It has widened but has not developed into an incipient ocean.

## 4.2.2    The geological framework

Figure 4.3 depicts the generalized geology of the African and Arabian Plates
adjacent to the Red Sea. From the data presented there and from your
existing knowledge, attempt Question 4.3.

> **Question 4.3** For both (a) and (b) below, complete the sentence by
> selecting the most appropriate statement from A–D. (Several inferences
> have to be made, so do not spend too long pondering over your answers.)
>
> (a)  The Red Sea basin developed …
>
> >   A    … during the Precambrian.
> >
> >   B    … between 250 and 550 Ma (during the Paleozoic).
> >
> >   C    … after the Permian (in the past 250 Ma).
> >
> >   D    … during the Tertiary (in the past 65 Ma).
>
> (b)  The rocks of north-east African and Arabian continental crust that now
>       flank the Red Sea are …
>
> >   A    … primarily igneous and metamorphic rocks formed 500–1200 Ma
> >        ago.
> >
> >   B    … ancient igneous and metamorphic rocks formed more than
> >        1200 Ma ago.
> >
> >   C    … sedimentary rocks formed during the Phanerozoic.
> >
> >   D    … recent igneous rocks.

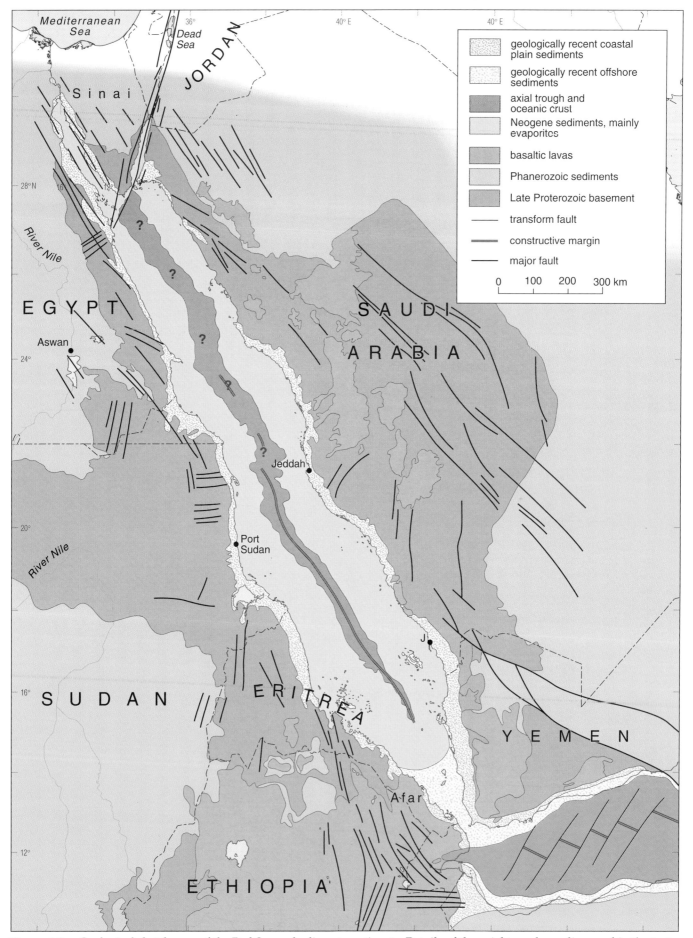

**Figure 4.3**   Geological sketch map of the Red Sea and adjacent continents. Details of the axial trough are discussed in the text. J = Jebel-at-Turf.

Down the centre of the Red Sea, an axial trough is picked out by the 500 m bathymetric contour. Figure 4.4 is a vertical cross-section through the upper crust of the southern Red Sea which shows the morphology of the axial trough.

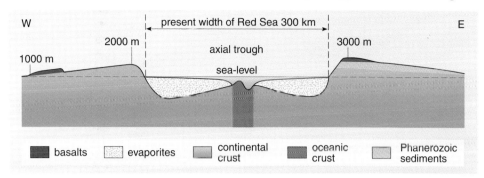

Figure 4.4 Idealized composite E–W section across the southern Red Sea and adjacent areas showing main rock units. Note that the details shown will not fit any single section on Figure 4.3 — the section has been drawn to show regional geological relations.

The trough is well-defined over much of the Red Sea, particularly in the south, and its bathymetry is, at first sight, very similar to that in other oceanic rift zones such as the Mid-Atlantic Ridge. Moreover, like the Mid-Atlantic Ridge the axial trough is flanked by sediments. However, there is no equivalent of a continental slope or rise in the Red Sea, for the shallow water gives way abruptly to the deep waters over the axial trough. This suggests we are not dealing with a continental shelf formed primarily of continental clastic detritus. Indeed, deep drilling during the exploration of the area for oil has shown that these marginal waters are underlain by about 4 km of Neogene (24–5 Ma) evaporites.

A further interesting observation is that escarpments are readily identifiable on both sides of the Red Sea — the Arabian upper crust is almost a mirror image of that on the African side of the axial trough. You should also note from Figure 4.4 the presence of basalts on both sides of the Red Sea. They erupted onto an erosional surface which is Mesozoic to early Tertiary in age and extend over a total area of about 400 000 km$^2$, reaching thicknesses of up to 2 km in Ethiopia. They are obviously Tertiary in age and radiometric dating shows the great majority of lavas to be 30 Ma or younger.

> Question 4.4   From the data presented so far, is there any evidence for magmatic/tectonic activity in the period between the formation of the crystalline basement c. 500–1200 Ma ago and the outpouring of the Tertiary lavas?

During the Mesozoic, the shallow seas advanced and retreated occasionally, leaving behind thin sedimentary horizons. It is only during the Cenozoic that anything of major magmatic and tectonic significance occurred. Geological investigations in Arabia and Ethiopia indicate that the first major phase of this activity started about 30 Ma ago with the outpouring of vast quantities of basaltic lava onto the low-lying erosional surface. We shall return to the interrelated causes of both volcanism and uplift later, but first we review the sedimentological evidence for the evolution of the Red Sea basin.

## 4.2.3   The sedimentary record of the Red Sea

Much detailed information can be obtained about the paleogeography of an area by studying the sedimentary record. The composition, fossil content and texture of sedimentary rocks reflect the depositional environment and, in the case of terrigenous sediments, the composition and relief of the source area(s). Coarse clastic sediments such as grits and conglomerates indicate high relief and a local sediment supply, whereas fine clastic sediments, such as muds and sandstones, are produced after longer periods of transport across areas of low relief. Evaporites form in basins in which the rate of water loss by evaporation exceeds water supply and are indicative of enclosed basins in hot climates. By contrast, limestones and corals form only in warm clear open seawater. In the case of the

Red Sea, the sedimentary record covers only the relatively recent geological past and during this time Africa has moved slowly through tropical and sub-tropical latitudes. This being so, no major shift from one climatic belt to another is involved. A detailed understanding of the sedimentary record is beyond the scope of this Block. However, the following is a review of the sedimentological history of the Red Sea, containing much useful information of the paleogeography of the region.

The distribution of marine sediments shows that, from 85–20 Ma, a warm shallow sea extended across the northern Red Sea and nearby continental areas, but the southern part of the present-day basin lay above sea-level until at least 40 Ma ago. Around 15 Ma, there was a dramatic change in the sedimentation which became restricted to the present extent of the Red Sea basin and the southern Red Sea became an area dominated by evaporite deposition.

The presence of evaporites indicates that during the mid-Neogene the Red Sea was an enclosed basin filled with brine that was periodically subjected to influxes of fresher oceanic water. A total accumulated thickness of 4 km of evaporites resulted. Given that it takes evaporation of 1 km depth of ocean water to produce ~3 m of salts, this must have required many flooding events. A similar record of evaporite deposition during the Neogene is found in the Mediterranean, further implying a direct link between it and the Red Sea. However, coral reef horizons are also found throughout the Red Sea floor sediments deposited between 15 Ma and 5 Ma. Coral reefs cannot survive in the very saline conditions required to precipitate evaporites and their presence is an indication of periodic flooding by seawater from the north. Thus, we know that the Red Sea basin was open to the ocean during parts of this period.

> Question 4.5 Bearing in mind that the Red Sea evaporite deposits reach thicknesses as great as 4 km, and evaporites form in shallow waters, what does this tell you about the evolution of the Red Sea basin?

Conglomerates, which indicate the rapid development of contrasts in relief and increased continental erosion, were first deposited on the Red Sea flanks at about 20 Ma and continue to form there up to the present day. Their high energy depositional conditions mark the onset of rising escarpments, caused by extensional tectonics, that dominate both flanks today, especially in the southern Red Sea.

Around 5 Ma ago, the end of evaporite deposition and the introduction of marine sediments with Indian Ocean fauna records a second major change in sedimentation. The faunal change is strong evidence that the Red Sea was breached from the south, allowing continual exchange of water between the Red Sea and the Indian Ocean. Moreover, the absence of sedimentary deposits in the Suez area suggests that at the same time the land rose to the north, creating a low barrier between the Red Sea and the Mediterranean.

So, the sedimentary record suggests the development of a basin during the Neogene (23.8–5.3 Ma). Continuous subsidence in the basin to a depth of 4 km occurred concurrently with the development of rising flanks and a growing escarpment on the seaward side of the flanks to present heights of 2–3 km above sea-level. Originally, this basin was open to the north but at 5 Ma the basin opened to the Indian Ocean in the south when the Red Sea became isolated from the Mediterranean.

## 4.2.4    Summary of Section 4.2

- The Red Sea lies in an elongate basin within late Proterozoic continental lithosphere which developed as a rift during the Tertiary.

- If Arabia is rotated clockwise from its present position relative to Africa, the Red Sea can be closed, but there are large areas of overlap both in the north and in the south.

- At the southern end of the Red Sea, a triple junction is formed from the intersection of the Red Sea Rift, the Gulf of Aden and the East African Rift.

- The southern Red Sea basin lies in a region of domal uplift and has widespread post-30 Ma basaltic magmatism exposed on both its uplifted flanks.

- The Red Sea basin is floored by 4 km of evaporites which provide evidence of subsidence in an enclosed basin between 20 Ma and 5 Ma.

- Flanking conglomerates in the Red Sea basin provide evidence for an escarpment developing since 20 Ma.

# 4.3    Basaltic magmatism in the Red Sea

## 4.3.1    Distribution of basaltic rocks in the Red Sea region

The distribution and age of basaltic rocks in and around the Red Sea Basin are fundamental to our understanding of its development. Unlike the Kenya Rift, the great majority of volcanic rocks are basaltic, although more evolved rocks are present. However, it is the basalts that tell us about the thermal structure of the mantle, so we shall concentrate in this Section on their distribution, composition and modes of origin.

The distribution of the different types of basalt are shown in Figure 4.5. In addition to the now familiar terms alkali basalt and tholeiite, there is also reference to so-called transitional basalts. These, as their name implies, have characteristics intermediate between alkali basalts and tholeiites. They plot close to the dividing line on the total alkalis versus silica plot and have either minor Hy or minor Ne in their normative mineralogy.

> **Question 4.6**  Is there any correlation between the amount of extension and the location of the different types of basalt? Comment on any associations that you have noted.

So, there is a general relationship between basalt composition and tectonic location but also some exceptions to this rule that you will explore later.

## 4.3.2    Regional chronology of basaltic magmatism

Basaltic magmatism in the Red Sea, Ethiopia, Yemen and the Gulf of Aden spans the past 40 Ma, although the detailed chronology of magmatic events is only now being unravelled with the use of high precision geochronology. The main technique employed in these modern investigations is a variant of the potassium–argon technique that you may have encountered in level 2 courses. This high precision technique is referred to as $^{40}$**Ar**/$^{39}$**Ar dating** and is described in Box 4.1.

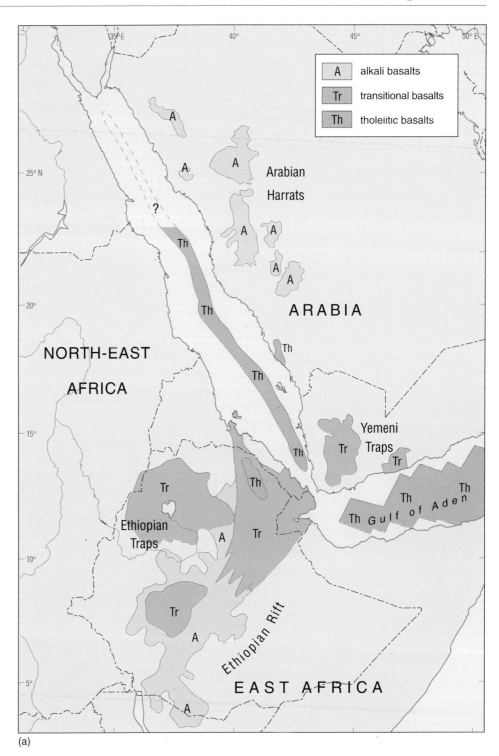

**Figure 4.5** (a) Distribution of basalt types in the Red Sea area. Note that the basaltic provinces are highly simplified and are characterized by the predominant basalt type. (b) Age provinces of basaltic magmatism in the Red Sea region.

(a)

## Box 4.1 $^{40}$Ar/$^{39}$Ar dating

$^{40}$Ar/$^{39}$Ar dating is a refinement of the K–Ar dating method, based on the β-decay of $^{40}$K to $^{40}$Ar. It has wide application in Earth sciences because potassium-bearing minerals, such as mica, feldspars and amphiboles, are found in many rock types. In the K–Ar technique, the concentration of elemental K in a sample is measured by chemical means while the abundance and isotope ratio of Ar is measured using a mass spectrometer. In the $^{40}$Ar/$^{39}$Ar method, the sample to be dated is first irradiated in a nuclear reactor to transform a small proportion of the $^{39}$K into $^{39}$Ar, through interaction with fast neutrons. $^{39}$Ar is radioactive with a half-life of 269 years and is not present in natural samples. Therefore, any $^{39}$Ar present in an irradiated sample must have been generated from potassium and the $^{39}$Ar signal measured in the mass spectrometer is a measure of the potassium in the original sample. So, in the irradiated sample, the ratio of $^{39}$Ar to $^{40}$Ar is related to the original $^{40}$Ar/$^{40}$K ratio and

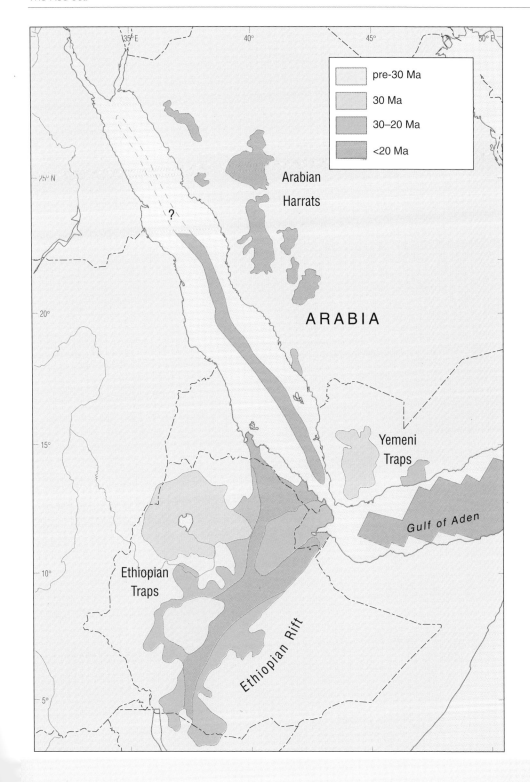

an age can be determined from one mass spectrometric measurement.

Argon is extracted from the irradiated sample by heating and eventually melting it in an ultra-high vacuum. There are two ways in which the sample can be heated. The first involves heating in a furnace and gradually increasing the temperature in increments until the sample melts. Isotope measurements are made at each temperature increment and there are numerous methods for dealing with the variations of ages derived from each step. A more recent development uses a laser to heat a small area of an irradiated rock or mineral and hence allow the determination of the age from the gas released from each of a number of laser shots. Both methods have been applied with equal enthusiasm to the dating of basaltic rocks.

The ages of basaltic rocks in north-east Africa and Arabia and the surrounding ocean basins are summarized in Figure 4.5b. The earliest basaltic activity in north-east Africa occurred between 45 and 35 Ma in southern Ethiopia and was of **transitional** composition, i.e. basalts with compositions very close to the CPSU with small amounts of either Ne or Hy in their norms. However, this event does not appear to be directly related to events in the Red Sea but is one of the earliest volcanic episodes in the evolution of the Kenya Rift and will not be considered further here. The first and most voluminous volcanic event in the evolution of the Red Sea was the eruption in northern Ethiopia of thick sequences of monotonous basalts that now make up the Ethiopian Highlands. Eruption of these so-called Ethiopian Traps coincided with the eruption of the Yemeni Traps on the Arabian peninsula and in places they are 2 km thick, making up the largest volume of continental basaltic volcanism in the region. They first erupted at 29–30 Ma and continued erupting in ever-decreasing volumes until 25 Ma. At the same time, transitional basalt dykes were intruded parallel to the present-day coast of the Red Sea and there is a particularly dense swarm along the Arabian coast at Jebel-at-Turf (Figure 4.3) dated at 29–30 Ma.

After this early episode, basaltic volcanism migrated to those areas affected by current extension, namely the Afar Depression and the Ethiopian Rift, while alkaline basalts started to erupt over the trap sequences of the Ethiopian Highlands. This activity began at 20–22 Ma and continued episodically throughout the Neogene and up to the present day.

The final phase of activity began about 10 Ma with the eruption of tholeiitic basalts in the Gulf of Aden followed by the Red Sea axial trough at 5 Ma. Once again, this activity continues to the present day. Throughout this time period, alkali basalt eruptions have occurred on the Arabian Peninsula, forming the laterally extensive flows of the Arabian Harrats. In the following Sections, we shall concentrate on different aspects of this volcanic history, beginning with the Ethiopian and Yemeni Traps.

### 4.3.3   The Ethiopian and Yemeni Traps

The locations of the Ethiopian and Yemeni Traps are shown in Figure 4.5. In places, the lava sequences are over 2 km thick and their present areal extent is about 400 000 km². However, considering the amount of erosion required to form the dramatic escarpments at the edges of the plateau (Figure 4.6), this is very much a minimum estimate. In addition to the two main plateau regions of Ethiopia and Yemen, there are a number of outliers that give some indication of the original areal extent of the basalts. A circle that includes most of these outliers covers an area of about 750 000 km² (Figure 4.7). Assuming an average thickness across the whole region of 1 km, this implies a total volume of 750 000 km³. (Note that the older basalts of southern Ethiopia have been excluded from this estimate.)

The remarkable feature of the Ethiopian Traps, as with other examples of such so-called **flood basalts** worldwide, is their uniformity. Figure 4.6 shows a view of the escarpment at the north-eastern edge of the Ethiopian plateau, revealing the 'layer-cake' stratigraphy of many thick monotonous flat-lying lava flows. There are relatively few intervening beds or layers of other lithologies and no angular unconformities between flows or groups of flows.

◉   What do you think that this stratigraphy implies?

◉   Little time for the development of other deposits and little tectonic activity between successive lava flows.

**Figure 4.6** Photograph of the escarpment at the northeastern edge of the Ethiopian flood basalts (or traps). The vertical cliff face is over 1000 m high and the total thickness of basalts exposed in this section is almost 2 km.

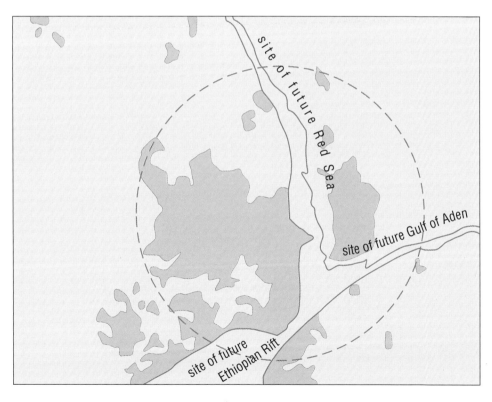

**Figure 4.7** Approximate reconstruction of the geography of the Red Sea and Gulf of Aden at the time of flood basalt eruption (~30 Ma). The Red Sea, Gulf of Aden, Afar Depression and Ethiopian Rift have been closed up. Basalt outcrops are shown in pink, older rocks in buff. Circle has a radius of 500 km.

Clearly, the Ethiopian/Yemeni Traps are a huge volcanic province and the monotonous stratigraphy suggests they erupted rapidly. But how long did it take to build up?

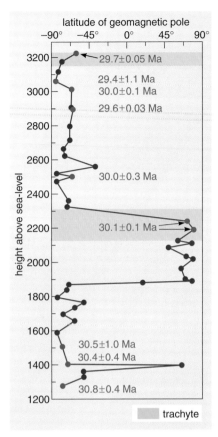

**Figure 4.8** A compilation of $^{40}Ar/^{39}Ar$ ages and paleomagnetic orientation of basalts in the 1950 m section illustrated in Figure 4.6. The latitude of the geomagnetic pole of successive samples is shown by the blue points joined by blue lines.

### 4.3.4 Geochronology of the Ethiopian Traps

Two techniques give estimates of the absolute age and duration of the Ethiopian Traps: $^{40}Ar/^{39}Ar$ geochronology and a method based on the timing of paleomagnetic field reversals (magnetostratigraphy). Figure 4.8 summarizes the results from one section that has been studied in detail. This section is through the thickest part of the sequence in Ethiopia (1950 m) and it comprises 43 basaltic flows with two trachytic tuffs, one in the middle of the sequence and the other at the top.

● Why do you think these two trachytic units are important for $^{40}Ar/^{39}Ar$ geochronology?

● They contain alkali feldspar which is rich in potassium, the parent of $^{40}Ar$.

Alkali feldspar is not only a major host of potassium in these rocks, it is also less susceptible than the groundmass to post-eruption alteration due to weathering. Hence, ages derived from separated alkali feldspar crystals are very precise.

● Where is the potassium located in basalt?

● Basalts are generally poor in potassium and rarely contain K-rich minerals. Any potassium is therefore concentrated in the groundmass that represents the residual liquid produced after the crystallization of the major phases, olivine, pyroxene and plagioclase.

Some of the potassium in basalts is located in plagioclase and analyses of plagioclase phenocrysts separated from basaltic hosts can give precise Ar ages. However, flood basalts are often aphyric (no phenocrysts) and so for most cases we rely on analyses of bulk samples. As the groundmass is frequently the first part of the rock to suffer from low temperature alteration processes, Ar ages of basalts are generally of poorer precision than those from more evolved rocks.

**Question 4.7** (a) Given the geochronological data in Figure 4.8, what is the maximum and minimum duration of basalt volcanism in this section? The ± figures are a measure of the uncertainty of each age determination and should be taken into account when answering this question. (b) Assuming the maximum duration of volcanism as calculated in (a), what is the mean interval between eruptions in this section?

The answers to these questions reveal both the geologically short duration of the whole flood basalt event and the long periods (on a human timescale) when little was happening.

#### Magnetostratigraphy

Studies of magnetic anomalies on the ocean floor reveal that the Earth's magnetic field oscillates between periods of normal and reversed polarity and that these oscillations occur on a timescale of tens or hundreds of thousands of years.

● How does this period compare with the estimated periodicity of flows in the Ethiopian flood basalts?

● It is very similar.

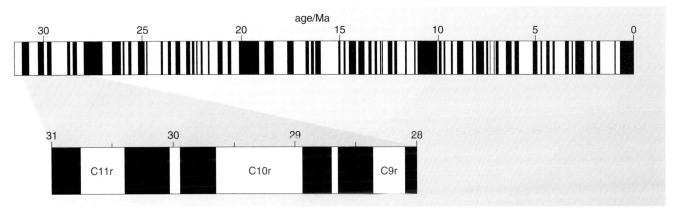

Figure 4.9   A calibrated magnetic reversal timescale for the past 32 Ma and an enlargement of the changes between 31 and 28 Ma. The dark bands represent periods when the Earth's magnetic field was normally polarized (i.e. as today with magnetic north close to the geographic North Pole). The light bands represent periods of reverse polarity. A chron is defined as a period of time which includes an episode of reversed followed by an episode of normal magnetization (i.e. a white followed by a black band), referred to by the letters r and n respectively. Thus, C10r refers to magnetic chron 10, reversed episode and C10n magnetic chron 10, normal episode. Brief reversals such as the one at 30 Ma are ignored when numbering the chrons.

Thus, flood basalt sections should record these magnetic reversals and so enable us to place further constraints on the timing and duration of volcanism. This is indeed the case with the Ethiopian flood basalts. The section shown in Figure 4.8 has a very simple magnetostratigraphy that is easy to interpret. The lower 600 m shows reversed magnetization, the central 300–400 m is normally magnetized and the top 900–1000 m is reversed. Of the three magnetic 'zones', only the central one is fully represented, i.e. both the bottom transition from reverse magnetization and the top transition back to reverse magnetization are included in the section. Fortunately, this contains one of the trachytic tuffs, precisely dated at 30.1 ± 0.1 Ma. This age can be compared with a calibrated reversal timescale derived from other locations, shown in Figure 4.9.

**Question 4.8**  (a) To which chron do you think the central normally polarized rocks of Figure 4.8 belong? (b) If this correlation is correct, what is the maximum duration of volcanism in the section in Figure 4.8? (c) How does this compare with estimates of the duration from Ar geochronology?

The agreement between the two approaches gives us confidence that we are indeed dealing with a geological event of only 1.5 to 2 Ma duration. This confirms the initial conclusions from simple field observations that flood basalts appear to be erupted over geologically brief periods, a conclusion reinforced when ages from across the Ethiopian Highlands are compared. In most cases, ages of between 31 and 29 Ma predominate with relatively few younger ages and none older. The conclusion is that perhaps 90% of the whole sequence erupted in this limited time period.

**Question 4.9**  (a) What is the eruption rate (in km$^3$ per year) of the combined Ethiopian and Yemeni flood basalts? (b) How does this value compare with modern volcanic provinces (Table 4.1)?

**Table 4.1**   Eruption rates in modern basaltic provinces related to mantle plumes.

| Hawaii | 0.16 km$^3$ yr$^{-1}$ |
| --- | --- |
| Iceland | 0.24 km$^3$ yr$^{-1}$ |
| Azores | 0.01 km$^3$ yr$^{-1}$ |
| Kenya Rift | 0.03 km$^3$ yr$^{-1}$ |

This high eruption rate is much greater than anything active today on the surface of the Earth. Similarly high and even higher eruption rates have also been calculated from other continental flood basalt provinces. You will be returning to this issue in the next Section when investigating the wider role of mantle plumes in continental break-up.

## 4.3.5    Composition of the Ethiopian flood basalts

As with our studies of the Kenya Rift, a good place to start with any basalt province is with the total alkalis versus silica diagram (Figure 4.10). The analyses reveal that the bulk of the basalts plot close to but generally below the alkaline–tholeiitic divide. Hence their designation as transitional. However, more systematic variations emerge from variations in other parameters. The most useful of these is a plot of $TiO_2$ against Mg# (Figure 4.11), which divides the samples into three distinct groups. These have been classified as low-Ti (LT) and two high-Ti (HT1, HT2) groups as shown.

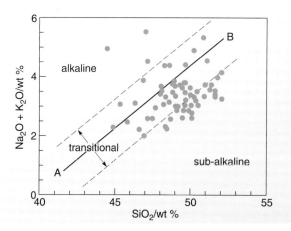

**Figure 4.10**  Total alkalis plotted against silica for the Ethiopian flood basalts. Line A–B is the alkaline–subalkaline divide from Figure 2.3.

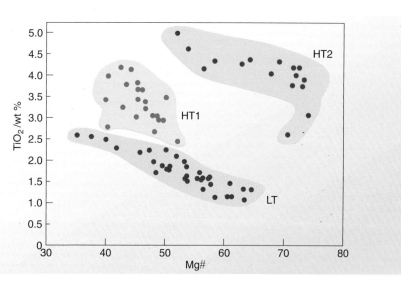

**Figure 4.11**  $TiO_2$ abundances in Ethiopian flood basalts, plotted against Mg#.

## Activity 4.1    Compositional variations in the Ethiopian flood basalts

Now attempt Activity 4.1, which explores the compositional differences between these groups of basalts. It should take you about one hour.

Your conclusions from Activity 4.1 indicate that the HT2 magmas are derived from depth by small degrees of melting of a garnet-bearing source region with trace element abundances similar to those of ocean island basalts. We have deduced from our study of eruption rates and volumes that a mantle plume must have been involved in the generation of the Ethiopian Traps and so it seems not unreasonable, based on this evidence, to relate the HT2 basalts to melting of the mantle plume.

By contrast, the LT basalts are derived from possibly larger degrees of melting at shallower depths of a garnet-free source with trace element abundances different from those of ocean island basalts. Either their trace element abundances have been modified before they reached the surface or they were derived from a source region distinct from any seen beneath the oceans. The debate concerning the source region of the LT basalts in Ethiopia has yet to be resolved but a strong possibility is that that they were derived from the mantle part of the continental lithosphere. This is based essentially on the difference between the LT basalts and oceanic magmas. Such a possibility raises a number of problems concerning the mechanism by which mantle lithosphere can melt and these will be addressed in more detail in Section 5.

The third group of basalts, the HT1 magma group, remains problematic. You can see from Figure 4.11 that they have $TiO_2$ contents intermediate between those of the LT and HT2 magma types and it is possible that they could represent a distinct magma type derived from yet another mantle source region. Alternatively, they could be a hybrid produced by mixing between the LT and HT2 magma types as their trace element abundance patterns are intermediate between the other two. Furthermore, their Mg# values are never greater than 50 and so there is no direct evidence that they have their own primary magma. Both of these observations suggest a hybrid or mixed origin. For now it is enough to recognize that within the Ethiopian CFB there are basaltic magmas derived from two distinct source regions, one that can be identified with a mantle plume and the second that is distinct from oceanic source regions and probably lies within the continental mantle lithosphere.

## 4.3.6   Basalts of the Afar Depression

The Afar Depression marks the junction between the continental rift system of East Africa and the incipient oceanic rift of the southern Red Sea (Figure 4.3). Topographically, the Afar is a triangle, delimited by huge escarpments (up to 2000 m high) capped by flood basalts to the south and west, and by the narrow neck of the sea to the east (Figure 4.12).

The geology of the area is dominated by basalts and it is to this area that the main focus of volcanism migrated after the eruption of the flood basalts in Yemen and northern Ethiopia. There is an almost unbroken record of basaltic activity from 29 Ma to the present day, interspersed with some rhyolitic and trachytic eruptions. The basalts are all transitional between true tholeiites and true alkaline basalts, and have trace element compositions that are very similar to ocean island basalts and hence to the HT2 flood basalts. It is this similarity that relates the Afar basalts to the underlying mantle plume, generally referred to as the Afar plume.

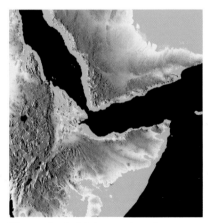

**Figure 4.12**  Major surface features of the Afar region at the southern end of the Red Sea.

> **Question 4.10**   You should recall from the introduction to this Section that Afar was one area where the coastal fit was poor. Given that the geology of Afar is dominated by basalts of <30 Ma, can you now suggest a reason for the poor fit?

### 4.3.7   Magmatism on the Arabian peninsula: dykes and harrats

Basalts of varying ages occur along the length of the western coast of the Arabian peninsula roughly parallel to the strike of the Red Sea. In particular, there are extensive lava fields, known as harrats (from the Arabic for hot), and dykes concentrated along the coastal plain. These occurrences are in addition to the Yemeni plateau lavas, the equivalent of Ethiopian flood basalts (Figure 4.5).

**Figure 4.13** Aerial photograph of the Chada flow, near Medina, Saudi Arabia, which erupted in 1256 AD. The black rocks are basalts, white areas are evaporite deposits from ponded water and the dun-coloured rocks are the basement.

The harrats have been divided into an eroded older series and a more extensive younger group. Together, they cover an area of $100\,000\,\text{km}^2$, but they are seldom thicker than 300 m. Hence, their maximum volume is about $3 \times 10^4\,\text{km}^3$, at least an order of magnitude smaller than the volume of the Ethiopian flood basalts. They are also dominantly alkaline in composition and any associated evolved lavas are alkaline phonolites.

Basalt dykes occur along the length of the southern Red Sea and are of alkaline or tholeiitic composition. The alkaline dykes are associated with the older harrats, and strike essentially N–S. Likewise, volcanic plugs associated with the older harrats lie along N–S alignments. The tholeiitic dykes are oriented more towards the north-west and are parallel to the Red Sea coast. They are also more numerous and of greater extent than the alkaline dykes, stretching from Jebel-at-Turf in the south to Sinai in the north (Figure 4.3), although their frequency declines northwards. Coast-parallel dykes also occur along the western coast of the Red Sea in Eritrea.

The dykes and older harrats have recently been dated using $^{40}\text{Ar}/^{39}\text{Ar}$ methods. These studies have revealed that the alkaline dykes associated with the older harrats are 27–29 Ma old whereas the tholeiitic dykes have ages between 21 and 24 Ma. Moreover, these ages show no systematic variation with location.

> **Question 4.11**   What does this sequence suggest about melting and extension through time?

> **Question 4.12**   Given the relationship between extension factor and melting (Figure 3.27) and assuming a mantle potential temperature of 1280 °C, at what value of $\beta$ does melting begin?

This analysis suggests that crust originally 35 km thick has to be thinned to about 9 km before melting begins. However, we know that at 30 Ma the Afar plume was already active to the south, producing the Ethiopian and Yemen flood basalts, so this value of $\beta$ must be regarded as a maximum as mantle of higher potential temperature would melt at lower values of $\beta$. Regardless of the amount of extension however, these dates show that extension across the area of the Red Sea must have started by 28 Ma and was well underway by 24 Ma.

Finally, the younger harrats have ages that range from 15 Ma to the present. Like their older counterparts they are alkaline in composition and many show a broken history of eruption in roughly the same place throughout their 15 Ma history. Their alkaline composition clearly implies a decrease in melt volume from the peak of activity associated with the tholeiitic dykes at 21–24 Ma to the present day. However, their continued activity implies thermal and/or tectonic disturbances within the mantle over this entire period.

### 4.3.8  Basalts of the Gulf of Aden and Red Sea axial trough

The basalts that have been dredged and drilled from both of these regions are invariably tholeiitic with low potassium contents. Chondrite-normalized rare earth element profiles of Red Sea and Gulf of Aden basalts are shown in Figure 4.14.

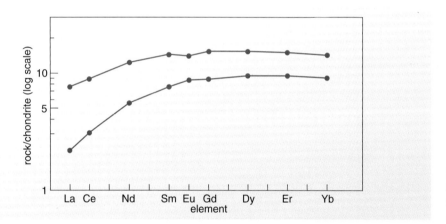

**Figure 4.14**  Chondrite-normalized rare earth element abundances of two representative basalts from the Gulf of Aden and the Red Sea axial trough.

🔵  Where else do similar basalts erupt?

🔵  The depletion of the light REE (La, Ce, Nd) in the basalts from the Red Sea axial trough is similar to that seen in many MORB (Figure 2.18).

The presence of low-K tholeiites in these areas demonstrates that extension has continued to such a degree that new ocean floor has been generated at a constructive plate margin. The ages of the ocean floor in these two areas can be determined from their paleomagnetic record, which we will deal with in the next Section.

### 4.3.9  Summary of Section 4.3

• Basaltic magmatism in the Red Sea Region started at ~30 Ma with the eruption of the Ethiopian and Yemeni flood basalts.

• The Ethiopian flood basalts are of transitional composition and erupted over a short period of time (1.6–2 Ma).

• The high-Ti (HT2) magma types were sourced in the Afar plume whereas the low-Ti (LT) magmas were derived from shallower levels, probably in the mantle lithosphere.

- The Afar Depression is a region of active extension, floored by transitional and tholeiitic basalts that date from 25–30 Ma to the present day.

- Basaltic magmatism on the Arabian peninsula comprises flows (harrats) and dykes that range in age from 27–29 Ma and 15–0 Ma.

- The harrats and oldest dykes are alkaline whereas younger, coast-parallel dykes are tholeiitic.

- Basalts from the Gulf of Aden and the Red Sea axial trough are low-K tholeiites (MORB).

## 4.4   Continental or oceanic lithosphere?

The Red Sea and the Gulf of Aden represent two excellent examples of the early stages in development of ocean basins by continental rifting. A central problem in understanding their evolution has been to determine how much of the extension during basin formation was accomplished by stretching of old continental crust, through faulting and crustal thinning, and how much by creation of new oceanic lithosphere.

Although there is general agreement that some oceanic crust has developed within the axial trough of the Red Sea, there has been considerable debate about the width and length of this area of oceanic crust. Many of the arguments result from extrapolating results from one east–west section of the Red Sea along its entire length, and it is becoming increasingly clear that the nature of the crust beneath the Red Sea differs from the north to the south. For this discussion, we shall define the southern Red Sea as that south of 21° N, the northern Red Sea as that north of 25° N, separated by the central Red Sea between 21° and 25° N.

### 4.4.1   The southern Red Sea

Figure 4.15 shows an aeromagnetic profile of magnetic anomalies across the southern Red Sea, approximately perpendicular to the spreading axis, and a depth profile along the same section.

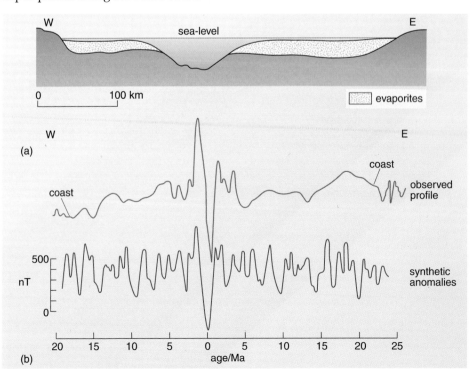

**Figure 4.15** (a) A sketch bathymetric profile across the Red Sea. (b) Observed and synthetic magnetic anomaly profiles (in nanoteslas) of the same section. The synthetic anomaly profile was calculated for a half-spreading rate of 1 cm yr⁻¹.

**Question 4.13**  Can you identify from Figure 4.15 three types of magnetic anomaly in terms of amplitude and wavelength? Is there any correlation between these anomalies and the bathymetric features shown in Figure 4.15a?

We shall use the terminology of type 1, 2 and 3 magnetic anomalies for the subsequent discussion. The type 3 anomalies are restricted to the continental crust and can be directly correlated with mafic dykes and intrusions, many of which are parallel to the Red Sea but do not exactly follow the trend of the axial trough.

Type 1 anomalies can be traced along the axial trough of the Red Sea for hundreds of kilometres and are generally parallel to the axial zone. They are of a wavelength and amplitude comparable with anomalies over mid-ocean ridges. Together with the composition of basalts dredged from the axial zone, they constitute powerful evidence that the axial trough is underlain by oceanic crust. But if these anomalies are caused by sea-floor spreading, it should then be possible to date the time of formation of the rocks producing them from the geomagnetic reversal timescale, in a similar way that we applied it to the Ethiopian flood basalts (Figure 4.9).

The shape and distribution of magnetic anomalies are good guides to the age of the oceanic crust producing them and the spreading rate at which this crust was formed. Given the spreading rate, the possible age range of the oceanic crust produced, the thickness of the magnetized layer and the strength of magnetization, then a synthetic magnetic anomaly can be calculated. Comparing synthetic anomalies with the observed anomalies leads to the best fit.

**Question 4.14**  Figure 4.15b shows the observed profile, and a synthetic profile calculated for a spreading rate of 1 cm yr$^{-1}$ per flank. Compare the synthetic and observed profiles. What can you deduce from the comparison in terms of distance, time, structure and spreading rate?

The interpretation of anomalies away from the axial trough has proven most controversial since their discovery during the early 1970s. Two competing interpretations fuelled heated debate amongst different research groups for many years. One group regarded them as resulting from the intrusion of mafic magmas into extended continental crust, but on a larger scale than seen on the flanks of the Red Sea. The second group proposed oceanic crust beneath the whole of the Red Sea, the magnetic anomalies reflecting spreading and crust formation at a time when reversals were less frequent but masked by the thick evaporite layers. The consensus now is that the interpretations of the first group are closer to the real situation. True sea-floor spreading in the Red Sea is restricted to the axial trough and has occurred during the past 5 Ma. The remainder of the Red Sea basin, beneath the thick accumulations of evaporites, is underlain by attenuated continental crust, intruded by linear mafic bodies that make a significant contribution to crustal extension. This interpretation implies two phases of tectonic activity: the first between 30 Ma and 5 Ma characterized by crustal extension and rifting; and the second from 5 Ma to the present in which true sea-floor spreading occurred.

## 4.4.2   The Afar Depression

Although the Afar Depression is floored by basaltic lavas, these are distinct in composition from those of the axial trough, being transitional rather than low-K tholeiites. Geologically, the crust beneath Afar is very different from true continental crust but neither is it typical oceanic crust. Figure 4.16 is a map showing the calculated depth to the Moho beneath Afar based on gravity and seismic data. In studying this diagram, remember that the Moho is, by

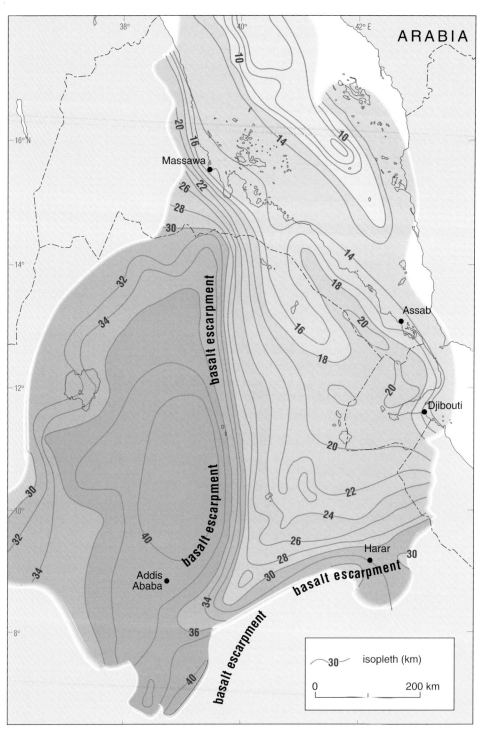

**Figure 4.16** Contour map of variations in depth to Moho in the southern Red Sea computed from gravity and seismic data.

definition, the base of the crust and that it is on average about 35 km deep under continents whereas beneath oceans it generally lies between 5 and 15 km beneath the sea-floor. In the Red Sea, allowing for varying water, evaporite and crustal thicknesses, the depth to the Moho lies between 11 and 15 km below the sea-floor.

**Question 4.15**   Study Figure 4.16 and attempt to answer the following questions. (a) What is the correlation between major topographic features (see Figure 4.12) and crustal thickness? (b) Is the crust of normal thickness under the Afar or has it been attenuated? (c) Are there any areas where continental crust might be absent?

Seismic refraction profiles from the same survey identified an upper crust with P-wave velocity of 6.1–6.2 km s$^{-1}$ across the Afar Depression, which is a higher average velocity than that measured in normal continental crust and is intermediate between the values we would expect from normal continental and oceanic crust. Moreover, the upper mantle beneath the depression has a velocity of 7.4 km s$^{-1}$ which is similar to that of asthenospheric mantle located beneath ocean ridges and too low for solid sub-continental lithosphere which lies in the range 7.8–8.4 km s$^{-1}$. Taken together, the geophysical data argue for severely thinned continental crust beneath Afar intruded by high P-wave velocity material. This would account for both a thin and a high seismic velocity 'continental' crust. The low velocity of the upper mantle suggests that asthenospheric mantle rises to shallow depths beneath Afar.

The geology of the Afar Depression, as we have noted above, is dominated by basaltic lavas and volcanoes. In places, the measured thickness of the flows reaches many kilometres and it is apparent that young basalts (<30 Ma) make up a significant proportion of the crust. Regional $\beta$ factors across Afar are estimated to be about 2 and locally they are as high as 6, almost high enough for the generation of oceanic crust. As with the Red Sea basin away from the axial trough, there are magnetic anomalies with strong linear trends but no regular striping as expected from true oceanic crust. Thus, the Afar is considered to represent a region of highly attenuated continental crust, intruded and thickened by basaltic magmas. It may be that the Red Sea basin beneath the evaporites has a very similar structure.

- If the Red Sea basin beneath the evaporites has a similar crustal structure to the Afar, why is the latter close to sea-level?

- An underlying mantle plume supports the whole region.

The amount of extension in Afar would normally reduce elevation to well below sea-level. However, the abundance of volcanic activity suggests the presence of unusually hot mantle beneath this region, hence the name Afar mantle plume. This name is, however, somewhat misleading. Basaltic magmatism is only located in Afar because this is the area of maximum extension. The hot plume material has spread out over a large area, supporting the whole Ethiopian plateau, and is not necessarily centred on Afar. An alternative name might be the Ethiopian plume as its geophysical and topographic effects are most pronounced in that region. However, the Afar plume is well established in the geological literature and will be used in this study.

## 4.4.3   The northern Red Sea

The northern Red Sea differs from the south in that there are no type 1 magnetic anomalies over the axial trough. Moreover, no basalts have been dredged from this region. Dykes do persist along the Arabian coast of the Red Sea and increase in frequency towards the sea. In the absence of stronger evidence, highly attenuated continental crust, intruded with numerous mafic dykes, is thought to persist beneath the whole of the Red Sea basin north of 25° N.

## 4.4.4   Extension of continental crust

The extension of continental crust is a requirement of all models for the development of the Red Sea and we have explored modes of crustal extension in Section 2 in association with the development of the Kenya Rift. Did the Red Sea begin as a structure like the Kenya Rift or were its early stages different? We can investigate the earliest stages of extension in the Red Sea in the Gulf of Suez in the far north of the region (Figure 4.3). This area has been unaffected by

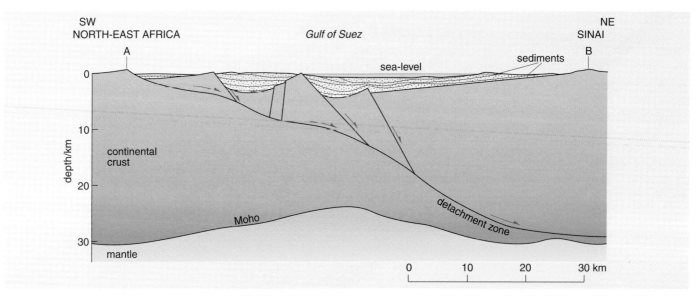

**Figure 4.17** A sketch vertical section across the Gulf of Suez based on seismic data.

magmatism because it is remote from the effects of the Afar mantle plume and also, as you will see, because the amount of extension is limited. Detailed seismic reflection profiles across the Gulf reveal the section summarized in Figure 4.17.

- Does the section in Figure 4.17 suggest a full or half-graben structure?

- Faulting and block rotation are asymmetric. The structure is therefore a half-graben.

The Gulf of Suez is a classic half-graben structure, with a major listric border fault and a number of antithetic faults that facilitate block rotation as shown.

> **Question 4.16** Estimate the amount of extension across the section in Figure 4.17 between points A and B. You will need to measure the present width between A and B and compare this with the estimated width when the four crustal blocks are rotated to their pre-rift positions. Express the amount of rifting as a beta factor ($\beta$).

The style of extension across the Gulf of Suez implies extension by simple shear, somewhat different from the current situation in the magmatically active Kenya Rift, but similar to that seen in the basins of the largely amagmatic Western Rift. However, as with both the Kenya and Western Rifts, the amount of extension across the Gulf of Suez is limited, in direct contrast to the amount of extension across the Red Sea basin just a few tens of kilometres to the south.

> **Question 4.17** Can you think of a reason for this?

We saw in the answer to Question 4.1 how moving the Arabian Plate south-westwards would completely close the Red Sea, except in the region of the Gulf of Suez where there was considerable overlap. Geological mapping in Sinai, Israel and Jordan has revealed a total of 105 km of sinistral movement along a fault known as the Dead Sea transform and seismic activity along its length suggests that it is still active. Thus, the movement required to generate the Red Sea basin can be accommodated by this structure. The details of when that movement occurred are uncertain but the offset of mid-Neogene dykes shows that movement began between 20 and 25 Ma. Reconstructions show that 30 km of movement occurred in the past 5 Ma, implying a slip rate of about 0.6 cm yr$^{-1}$.

To investigate the question of when extension occurred more precisely, we shall examine aspects of the geology in more detail and introduce a new technique that allows us to place some important constraints on the evolution of the Red Sea as a whole.

## 4.4.5    Summary of Section 4.4

- The southern Red Sea comprises a broad basin with a deep axial trough.

- The axial trough is floored by basalts and has large magnetic anomalies; it is true oceanic crust.

- The rest of the Red Sea is floored by evaporites, beneath which is attenuated continental crust, intruded by mafic dykes.

- The Afar Depression is floored by basaltic rocks with a composition distinct from those of the Red Sea axial trough.

- The crust beneath the Afar Depression is attenuated continental crust, intruded and thickened by basaltic intrusion.

- The northern Red Sea does not possess a well-defined axial trough and is attenuated continental crust throughout, again with some dyke intrusion.

- The Gulf of Suez is a half-graben in continental crust; amounts of extension are low.

# 4.5    The timing of extension in the Red Sea and Gulf of Aden

## 4.5.1    Geological evidence

Dating of extension and rift development can be achieved in a number of ways. The most direct approach relies on detailed studies of faults and the ages of rocks affected by their movements. In the case of the Red Sea, there are many normal faults that cut through the basement and overlying rocks on both margins. One area studied in some detail is in Eritrea, where the geology is relatively simple. Late Proterozoic igneous and metamorphic rocks are overlain by Mesozoic sandstones which are in turn overlain and overstepped to the north by a laterite soil horizon. The laterite is also present in Yemen and in Ethiopia and is in turn overlain by the flood basalts of the Ethiopian Trap sequences.

Laterite is a thick red soil formed in tropical climates by chemical weathering of silicate rocks. The weathering process is particularly intensive and involves the leaching of soluble and partly soluble oxides, leaving an oxidized residual soil enriched in red ferric oxide and aluminium oxide. It is this red coloration that is so typical of tropical soils. Laterites form in regions of minimal relief, otherwise they would be removed by erosion. So, the presence of laterite implies a widespread and lengthy period of humid weathering, and a regime with low relief and little sediment input, probably at or close to sea-level.

> Question 4.18    Figure 4.18 is a section through this geological sequence in Eritrea. (a) What are the possible minimum and maximum ages of the laterite? (b) The section also shows a series of faults that run parallel to the Red Sea coast. What is the maximum age of these faults? (c) Do the faults indicate extension or compression?

Turning now to sediments within the Red Sea basin itself, you read in Section 4.2.3 that much of the floor of the Red Sea is overlain by evaporites.

⬤    What is the age of these evaporites?

⬤    The evaporites on the Red Sea floor are of Neogene age.

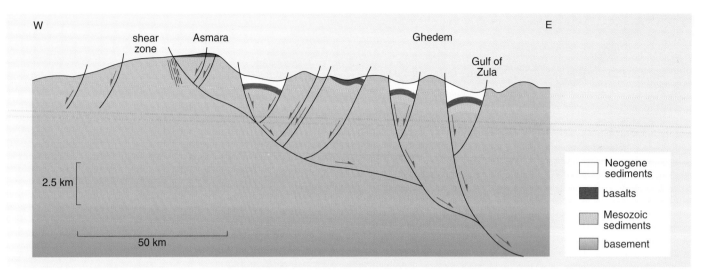

**Figure 4.18** A sketch section through the escarpment and coastal plain of Eritrea showing the location of faults and the displacement of the laterite and basalt horizons (shown as red).

The observation that the evaporites are restricted to the Red Sea basin is strong evidence that rifting had occurred before Neogene times.

Putting these two lines of geological evidence together tells us fairly precisely when rifting occurred. Normal faulting must post-date the flood basalts, i.e. <29 or 30 Ma ago. The Neogene began at 24 Ma but continued until 1.8 Ma. However, fossil corals within the evaporites suggest ages of 15 Ma. Therefore, the basin was developed enough to allow the influx of fresh seawater by 15 Ma, which suggests that the basin formed between 29 and 15 Ma.

The laterite horizon can also be used to measure the topographic effects of extension and plume emplacement over the past 30 Ma. Because laterites define a roughly horizontal datum that originally formed at or close to sea-level, their present altitude can be used to measure changes in elevation in response to extensional faulting. You should recall from Section 3 that extensional faults lead to subsidence of the hanging-wall to form the graben or half-graben basin, and uplift of the footwall. Both of these effects can be recognized in the coastal section in Eritrea.

On the coast of Eritrea, the laterite horizon outcrops between 75 and 175 m above sea-level. One hundred kilometres inland, it is found at a maximum altitude of 2400 m, giving an overall difference of 2 km. However, one particular block bounded by normal faults has a maximum altitude of 2750 m yet the surface is composed of relatively unweathered basement rocks with no laterite cover.

● What does the absence of laterite on this fault block suggest about the effects of faulting?

● The laterite and any weathered basement rocks have been removed by erosion. Their absence implies that locally fault movements must have produced uplift of at least 2750 m.

Thus, we can identify almost 3 km of vertical fault movement within 100 km of the Red Sea coast.

A further 150 km inland, beyond the most intense zone of normal faulting, the laterite is at a lower altitude but still at 500 m above sea-level. Away from the Red Sea margins and across much of the Ethiopian plateau, wherever the laterite is present it outcrops at this and even higher altitudes. The whole region has been uplifted following laterite formation close to sea-level.

● What is the likely cause of the regional uplift?

● Dynamic support from the underlying mantle plume.

Therefore, the base level of the surface of the continent in this region away from areas affected by faulting is around 500 m.

The development of rift basins consequently involves both upwards and downwards movement. The central graben subside, and in this case that subsidence is at least 500 m, while the flanks flex upwards. This phenomenon is known as flexural rebound and here amounts to an overall vertical displacement of between 2 km and 3 km, producing the striking escarpments along the length of the whole basin.

The above example shows how some constraints can be placed on the timing of extension and the throw of extensional faults where a dateable horizon is offset by those faults. However, in other areas, particularly in the northern Red Sea, the geology is less helpful in that easily dated basalts and the laterite are absent and normal faults are found within the basement where their displacements cannot be judged. In these regions, we have to resort to an alternative technique. One that has proven particularly useful in unravelling the evolution of the Red Sea is **fission track dating**.

## 4.5.2    Fission tracks and their application to Red Sea tectonics

The study of fission tracks provides a method for estimating the time at which a given sample cooled below a temperature of about 80 °C. Such temperatures are found at relatively modest depths in the Earth's crust (1–3 km) and so fission track ages record the time when a sample reached shallow depths, immediately prior to their exposure at the Earth's surface. Surface exposure of a rock originally at depth in the crust requires the combined effects of two processes.

● What are the two processes?

● Uplift and erosion of the overlying rocks.

However, erosion only occurs in areas where there is a topographic contrast. Think, for example, of scree slopes on the sides of mountains. In topographically high areas with little contrast in relief (i.e. plateaux), erosion is very slow. High elevation alone is not enough to expose a rock at the surface.

You have read how there are two components to uplift in the Red Sea region: dynamic support from the underlying Afar mantle plume and flexural rebound caused by extensional tectonics.

● Which of these two causes of uplift is most likely to be recorded by fission track ages?

● Flexural rebound, because it generates a marked contrast in relief — a fault scarp.

Fission track ages therefore give a good indication of the onset of extension, and in the following Sections two studies are highlighted to illustrate how fission track data can be interpreted and what they tell us about the timing and style of extension across the Red Sea.

## Box 4.2  The fission track technique

As you know from the methods of isotopic dating, many minerals contain small quantities of radioactive isotopes. Some unstable heavy elements such as $^{238}U$ break up by spontaneous fission into two lighter nuclei. These ejected particles damage the structure of the host mineral, leaving short tracks in their path which are a few microns long (Figure 4.19). The number of tracks in the mineral depends first on the quantity of parent isotope present and secondly on the time elapsed since the tracks were first preserved.

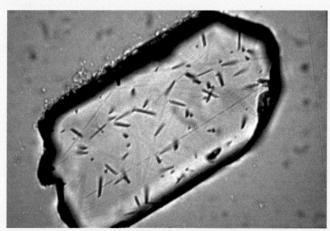

**Figure 4.19**  Uranium fission tracks in an etched apatite crystal. Photomicrograph is 1 mm wide.

The minerals usually used for fission track dating are apatite, zircon and sphene, because these all contain U. These are separated from the crushed rock and analysed for both the concentration of uranium and the number of fission tracks per unit area. In fact, of the few isotopes found on Earth that are subject to spontaneous fission, only $^{238}U$ has a short enough half-life and is sufficiently abundant to be of use for fission track dating.

An important point here is that at high temperatures these fission tracks are *not* preserved due to the ductile properties of crystals when heated. Fission tracks in zircon for example close up at temperatures above 200 °C and those in apatite close above about 125 °C. The number of tracks per unit area therefore results from the time elapsed since the mineral cooled below its **closure temperature**, and this is called the **cooling age**. If we analyse the uranium content of an apatite crystal, and count the track density within the apatite, a cooling age can be calculated. This is because the number of tracks results from both the quantity of uranium present, and the time elapsed since the tracks were preserved. A cooling age of 50 Ma indicates that the apatite was last at 125 °C (the closure temperature) 50 Ma ago.

In addition, extra information can be gleaned from the length of fission tracks. In essence, if a mineral is held at a temperature just below its closure temperature for a period of time, the damage to the crystal is repaired, but at a much slower rate than at higher temperatures. This is known as **partial annealing** and leads to tracks of variable length. In practice, the zone of partial annealing in which tracks are imperfectly recorded for geologically significant periods of time covers the temperature range from 80–125 °C. Thus, in addition to counting the tracks, the fission track scientist also needs to measure the length of each track — a painstaking task requiring a good microscope, a keen eye and the utmost patience.

### Yemen

In Yemen, in the south-west corner of the Arabian peninsula, fission track ages show a systematic distribution over a broad region shown in Figure 4.20. All the samples are derived from the late Proterozoic Basement, yet their ages range from those in excess of 300 Ma down to <25 Ma. Note how the oldest ages are from samples located furthest away from the Red Sea, and the younger ages are closest to the margins.

**Figure 4.20**  The geographical distribution of fission track ages of basement samples from Yemen.

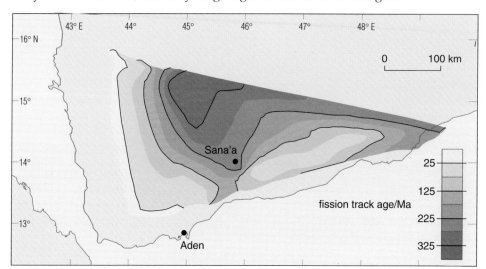

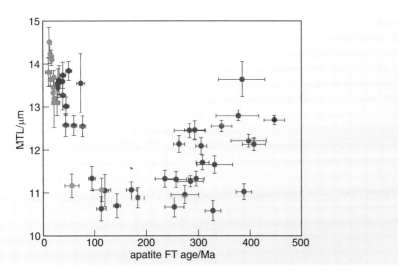

**Figure 4.21** The relationship between fission track (FT) age and mean track length (MTL) from basement samples in Yemen. Blue circles are data from samples from the Red Sea coastal escarpment, red circles are from inland or the coast of the Gulf of Aden. (1 μm = $10^{-6}$ m.)

Figure 4.21 shows the relationship between fission track cooling age and the mean track length of each sample.

⬤ Study Figure 4.21. Is there any relationship between the fission track (FT) age and the mean track length (MTL)?

◗ Yes, there is. Samples with the youngest ages, particularly those with ages <40 Ma, have the longest track lengths, those with intermediate ages (100–200 Ma) have short track lengths and those samples with ages >200 Ma have variable track lengths.

⬤ Focusing on those samples with ages <40 Ma and long track lengths, can you suggest a possible thermal history?

◗ These samples have resided in a warm environment, above the closure temperature, until ~40 Ma ago, after which time they were cooled rapidly, allowing the preservation of fission tracks.

Prior to 40 Ma, tracks produced in these samples were rapidly annealed and so not recorded. At about 40 Ma, the rocks were cooled rapidly to temperatures below 80 °C (below the partial annealing zone), allowing any tracks formed subsequently to be recorded

The second group of samples have ages >200 Ma and moderate-to-long tracks.

⬤ Again, can you describe the thermal history of these samples?

◗ These samples have been at low temperatures, i.e. at or close to the surface for >200 Ma. They have lain undisturbed for a long time and do not appear to have been affected by the event that brought the first group of samples to the surface.

Finally, there is a third group of samples with, on average, short tracks and ages intermediate between the other two groups. These samples have resided within the partial annealing zone for a considerable period of time and were brought to the surface during the later thermal event. Those tracks formed before ~40 Ma are short because they have been partially erased whereas those formed after 40 Ma are longer because of a lack of annealing.

This distribution of track lengths and ages is typical of many studies and is sometimes referred to as a 'boomerang' curve, for rather obvious reasons. This shape is shown schematically in Figure 4.22. The detailed analysis of this type of data is complex and also involves an analysis of how variable the lengths are in individual samples. However, the conclusions drawn from this study can be applied generally.

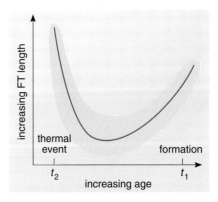

**Figure 4.22** Theoretical distribution of track lengths against age in a group of samples formed at time $t_1$ and cooled at time $t_2$.

In the case of the Yemen samples, the fission track data indicate initial cooling of the basement between 500 and 300 Ma and then a period of rapid cooling, caused by uplift and erosion since 40 Ma. The cluster of fission track ages at <40 Ma is a strong indication of the age of extension in the southern Red Sea. However, even within this group there is added complexity. Samples derived from the Red Sea margins (blue symbols in Figure 4.21) have ages of 20–25 Ma, whereas samples from the regions bordering the Gulf of Aden have older ages of between 30 and 40 Ma. These two ages from geographically grouped samples indicate that rifting occurred at an earlier time (Paleogene) in the Gulf of Aden than in the southern Red Sea (Neogene).

●  How does the estimate of the timing of extension from fission tracks in Yemen compare with results from field geology in Eritrea?

●  Very well. The fission track age of 20–25 Ma is included in the bracketed age of between 29 Ma and 15 Ma from field geology. However, there is no sedimentological evidence for rifting at an earlier stage.

Thus, extension in the southern Red Sea is constrained to have begun during the early Neogene, but how does that compare with the northern regions? This is important because it allows geologists to distinguish between alternative kinematic models for early rifting as outlined in Figure 4.23. In the first case, the African and Arabian Plates act as rigid blocks, both rotating about their own poles of rotation. Extension is synchronous along the whole length of the Red Sea and the plates do not undergo internal deformation. By contrast, the second model invokes propagation of the rift from, in this case, south to north. This model is similar to that suggested for the Kenya Rift (except that the latter propagated from north to south) and requires internal deformation of the two blocks as a result of the differential motion between the rifted and unrifted regions. In this case, fission track dates from the north of the Red Sea should give younger ages than those from the south.

## Northern Red Sea

The fission track ages of samples from the northern Red Sea are illustrated in Figure 4.24. For the sake of clarity, only those samples with ages <120 Ma are shown, although as with the Yemen data set, ages range up to 500 Ma and track lengths vary between 10 and 15 μm. Samples were obtained from the Egyptian coast of the Red Sea, north of 20 °N, and into the margins of the Gulf of Suez.

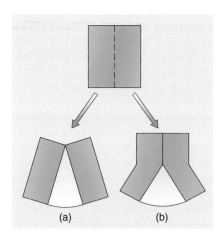

**Figure 4.23**  Possible kinematic models for the opening of the Red Sea. A crustal block fails along the dotted line as shown in the top part of the diagram. In (a), each block behaves rigidly with a 'hinge point' at the north. In (b), progressive migration of extension from south to north requires internal deformation of the two continental blocks.

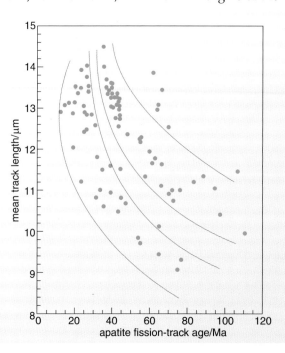

**Figure 4.24**  A plot of fission track age against mean length for samples from the northern Red Sea (1 μm = 10⁻⁶ m).

The data reveal not one but two cooling trends. The youngest of these terminates at 20 to 25 Ma and includes samples from both the Gulf of Suez and the Red Sea, whereas the better-defined trend terminates at 30–35 Ma and is dominated by samples from the Red Sea margin. The former age is similar to that derived from the Yemen data, confirming that a major extensional episode occurred in the Red Sea during the early Neogene (20–25 Ma). However, the older, Paleogene age was only poorly defined by the Yemen data and then only from samples along the Gulf of Aden. These new analyses first confirm the existence of an extensional episode during the Paleogene and show that it affected the northern Red Sea to a similar extent to the Gulf of Aden. The primary conclusion of these data is therefore that two episodes of uplift and hence extension are recorded along the length of the Red Sea. The first occurred in the Paleogene (30–35 Ma) and affected both the Red Sea and the Gulf of Aden. The second occurred around the early Neogene (20–25 Ma), but is recorded only in the samples from the Red Sea.

> **Question 4.19** Which of the two kinematic models of plate motion, illustrated in Figure 4.23, do these data support? Why?

There is therefore no sense that the Gulf of Aden rifted before the Red Sea, or that a particular part of the Red Sea rifted before another. Rifting did not propagate in one direction or another, as appears to be the case in the African Rift (Section 2). This simultaneous splitting of the Arabian Plate from Africa has implications for the nature of the driving forces behind rifting and drifting and these are the subject of the next Section.

### 4.5.3    Summary of Section 4.5

* Fault displacement of the laterite horizon in Eritrea reveals subsidence of the Red Sea basin and flank uplift caused by flexural rebound during the early Neogene.

* Sedimentary evidence indicates formation of the Red Sea Rift between 29 and 15 Ma.

* Fission track evidence points to two episodes of rifting, uplift and denudation, at 30–35 Ma and 20–25 Ma.

* Both events are recorded in both the northern and southern Red Sea although only the older is recognized in the Gulf of Aden.

* The data suggest simultaneous rifting along the length of the Red Sea.

* Both Arabia and Africa acted as rigid blocks during rifting.

## 4.6    Plate kinematics

The study of the geometry of plate boundaries and of the relative velocities of plate movements is known as **plate kinematics**. The location of plate boundaries is defined by the distribution of earthquake epicentres and we can use this criterion in the Red Sea region. Figure 4.25 shows the seismicity pattern in the Middle East. The broad zone of epicentres passing through the eastern Mediterranean and Asia Minor marks the destructive margin between the Eurasian Plate to the north and the combined Afro-Arabian Plate to the south. A distinct line of epicentres extends from the north-west Indian Ocean, through the Gulf of Aden where it forks with one limb running south-west down through Ethiopia and East Africa, and the other north-west along the centre of the Red Sea. As all these earthquake foci are shallow, of the type that occur at constructive plate margins, it appears that they divide the region into three plates known as the

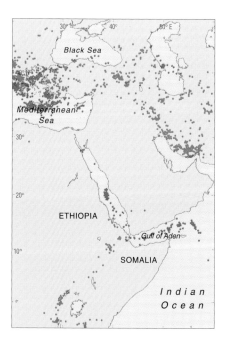

**Figure 4.25** The distribution of seismic epicentres in the Middle East.

African, Somalian and Arabian Plates, which meet at the triple junction between these plate boundaries, near the southern end of the Red Sea.

In the following discussion, we shall consider the movement of these lithospheric plates and the tectonic boundaries which separate them: the Gulf of Aden, the Dead Sea transform, the Red Sea constructive margin, the Suez Rift, and the East African Rift. Note that the Somalian 'Plate' is not strictly a distinct plate from Africa as continental lithosphere extends across the boundary, but it is useful to consider it as such in this discussion because relative movement between the Horn of Africa east of the East African Rift and the rest of Africa has resulted in the East African Rift system.

Movement between any two lithospheric plates can be described by a **pole of relative plate rotation** and the angle of rotation about that pole. The precise location of the pole and value of the angle of rotation requires the integration of geophysical and geological data and is constrained by the geometry of two rigid shells on a spherical surface. Considerable agreement exists regarding the plate motions during the past 4–5 Ma in the Red Sea region when sea-floor spreading is recorded by the magnetic anomalies in the southern Red Sea and in the Gulf of Aden. The average rates of relative plate motion based on data covering the past 5 Ma are given in Figure 4.26 and discussed in more detail in the following Sections. The older history of rifting back to the early Neogene and possibly the Paleogene is still debated, although in general the early plate motions appear to resemble the more recent movements.

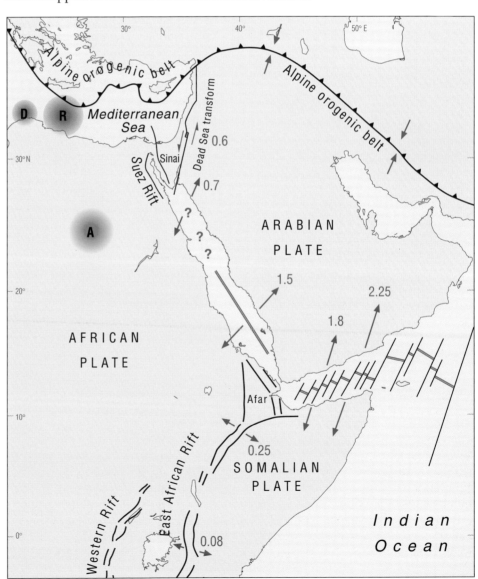

**Figure 4.26** Plate tectonic framework of the Red Sea area. Also shown are the approximate locations of poles of rotation for Red Sea (R), Dead Sea transform (D) and Gulf of Aden (A). Average rates of relative plate motion (in cm yr$^{-1}$) over the past 5 Ma are also given.

### 4.6.1    Gulf of Aden

The Gulf of Aden is bisected by a mid-oceanic ridge separating the Arabian and Somalian Plates. The ridge is offset by transforms which lie on small circles about the pole of rotation between these plates, which is located in northern Egypt around 25° N 25° E (Figure 4.26). Magnetic anomalies have been correlated to 10 Ma and possibly 14 Ma and indicate an angle of rotation of 4°–5° over the past 10 Ma. This corresponds to about 400 km opening in the central Gulf, or is equivalent to an average rate of relative plate motion of about 2 cm yr$^{-1}$ between the Arabian and Somalian Plates.

### 4.6.2    The Dead Sea transform

As suggested in Question 4.1, a close fit between Africa and Arabia is only possible at the northern end of the Red Sea if there has been movement along the Gulf of Aqaba.

> **Question 4.20**    Return to the reconstruction of Question 4.1. After closing up the Gulf of Suez, how much horizontal movement is necessary along the Gulf of Aqaba to make Africa and Arabia fit together?

A total movement of about 105 km along the so-called Dead Sea transform fault has been identified from matching lithologies and structures south of Lebanon, and seismic activity along the fault suggests it is still active. The Dead Sea transform, like all transform faults, defines a small circle with a unique pole of rotation. The present pole is located at 33° N 23° E in the southern Mediterranean, and about 30 km of left-lateral slip have occurred about this pole over the past 5 Ma. This yields an average slip-rate of about 0.6 cm yr$^{-1}$ (Figure 4.26).

Earlier movement, which totals 75 km (105 minus 30 km) of left-lateral slip, occurred about a pole a few degrees west of the present pole. The beginning of movement is constrained to lie between 24 and 20 Ma by the observation that in eastern Sinai both early Neogene dykes and distinctive structures in the basement are offset to the same extent. This suggests an average slip-rate a little lower than in the period since 5 Ma.

### 4.6.3    The Red Sea constructive margin

We have already used the seismicity pattern, basalt composition and magnetic anomalies to identify the Red Sea as an active plate boundary, but we can also determine the direction of plate motion due to any one earthquake by determining its fault-plane solutions. Such studies confirm that the Red Sea is a constructive boundary along which plates are moving apart along small circles which trend NE–SW where they cross the Red Sea axis.

> **Question 4.21**    Check back to Figure 4.1. Mark points on opposite coasts that coincide when a coastline fit is made. Does the spreading direction given by the reconstruction of Figure 4.1 agree with NE–SW spreading direction?

Young sea-floor spreading in the southern axial trough is recorded by type 1 magnetic anomalies of up to 5 Ma, with an average rate of relative motion of 1.5 cm yr$^{-1}$, producing about 75 km of opening by creation of new oceanic crust in the southern Red Sea. The summation of the movement along the Dead Sea transform and the Suez Rift suggests a total of 35 km in the northern Red Sea over the past 5 Ma, so the greater spreading rate to the south argues for a pole of rotation to the NNW. Though much studied, the Red Sea extensional boundary has not yielded precise information about the location of its pole of rotation because well-defined transform faults are absent. It is likely though that the pole lies in the eastern Mediterranean as indicated in Figure 4.26.

### 4.6.4   The Suez Rift

The rift developed in the Gulf of Suez is a classic example of extension of continental lithosphere through asymmetric thinning (Section 4.3.4). All studies of the region confirm that the Red Sea and Gulf of Suez rifted at about the same time. Drill holes and exposures in the rift show virtually no pre-Neogene conglomerates and so the geological evidence is against rifting before 25 Ma. This would imply that as soon as the Dead Sea transform was initiated, significant extension across the Gulf of Suez ceased. Other plate reconstructions imply that extension has continued across the Gulf of Suez and rifting has been simultaneous with strike–slip movement on the Dead Sea transform until the present day. There is no doubt that the Suez Rift remains active to some extent as can be detected from sporadic seismicity due to normal faulting and numerous faults which displace Quaternary sediments.

### 4.6.5   The Ethiopian Rift

The Ethiopian Rift is an integral part of the East African Rift system and its structure and magmatic evolution are similar to that of the Kenya Rift. At its broadest, where it joins the Afar Depression, it is about 100 km wide with $\beta$ factors up to 1.5 or 2. Further south, it narrows and $\beta$ factors reduce to values of 1.1 in southern Ethiopia. Offsets of lavas indicate extension began about 15 Ma ago and magmatism and shallow seismic activity continue to the present day.

### 4.6.6   Causes of rifting and drifting in the Red Sea

The important difference between the history of the East African Rift system as a whole and the Red Sea and the Gulf of Aden is that the latter two developed into constructive plate margins with ocean spreading systems producing low-K tholeiitic basalt and new ocean lithosphere. The question we now pose is why the East African Rift remained as the **failed arm** of a triple junction, while the other two arms continued developing into constructive margins?

We have shown that the Kenya Rift is probably a result of intraplate deformation. Extension is gravity-driven by stresses generated across areas of high relief and high heat flow, caused by the pull of old, cold and dense oceanic lithosphere immediately surrounding the continent. This process operates in the modern-day African Plate because it is almost completely surrounded by constructive margins and is isolated from the major driving force of plate motions, slab-pull. However, this was not the case during the early Tertiary, which is the time to which we must return in order to investigate why the Red Sea rifted in the first place and why it continues to spread today.

Prior to the initial rifting of the Red Sea at 30–35 Ma, the African and Arabian Plates were united in the Afro-Arabian Plate. Plate motion studies have shown that this plate was moving slowly towards the NE with a pole of rotation close to the Canary Islands.

●  What is the most direct method of determining absolute plate motion?

●  Measuring the rates of migration of basaltic volcanism along 'hot-spot' trails.

When a mantle plume impinges on the base of the oceanic lithosphere it can produce surface volcanism that forms intraplate seamounts and oceanic islands. If it is assumed that mantle plumes are fixed relative to the moving plates, they define a reference frame from which absolute plate motions can be derived. Thus, the rate of migration of basaltic volcanism along the Hawaiian chain reveals that the Pacific Plate moves to the west at a rate of about 70 mm yr$^{-1}$. The same technique has been applied to hot spot trails on the African Plate.

The clearest hot spot trails on the African Plate are defined by the Saint Helena and Tristan/Gough mantle plumes. Although there are a number of mantle plumes impinging on the base of the African Plate, those in the Atlantic Ocean, such as the Canary Islands and the Cape Verdes, are too close to the pole of rotation and to the African coast to provide well-defined trails. Only Saint Helena and Tristan/Gough in the Atlantic Ocean and Réunion in the Indian Ocean have produced volcanic trails that can be used as reliable indicators of plate motion.

Now attempt the following Activity, which explores the plate motion of Africa over the past 80 Ma.

## Activity 4.2    The absolute motion of the African Plate

Activity 4.2 involves determining the rate of migration of mafic magmatism along the Tristan hot spot trail in the Atlantic Ocean, and should take you about one hour.

This discovery of the slowing down of the African Plate is a relatively recent one, and is based on high precision Ar/Ar ages on samples collected during the 1990s. However, the slower velocity is also consistent with the ages of young seamounts along the St Helena hot spot trail and is solid evidence that the African Plate has been moving at a slower rate since about 30 Ma.

● What was happening in the Red Sea region at about this time?

● This age broadly coincides with the first phase of extension deduced from the fission track ages from the flanks of the Red Sea and also the eruption of the Ethiopian flood basalts.

The coincidence between these events implies that Red Sea rifting and the reduced velocity of the African Plate could be related. But why should this be? To answer this, we must look at plate boundaries, not of the African Plate, but of the reconstructed Afro-Arabian Plate, as shown in Figure 4.27. Many of the boundaries of this larger plate are the same as the African Plate, except that now the north-eastern boundary is marked by a destructive margin beneath the present-day Zagros mountains in Iran. During much of the Tertiary and back into the Cretaceous, this plate boundary marked the location of subduction of the so-called Tethys Ocean beneath the Eurasian continental landmass. It was the slab-pull exerted by this process that provided a significant proportion of the plate driving force for the north-east motion of the united Afro-Arabian Plate. At about 30 Ma ago, the rifting of the Red Sea isolated the African Plate from the Arabian Plate and hence from this important driving force. Since that time, the African and Arabian Plates have been behaving as independent plates, moving around different poles of rotation. The current absolute motion of Africa is now about a pole of rotation located in the eastern Atlantic Ocean, within the plate

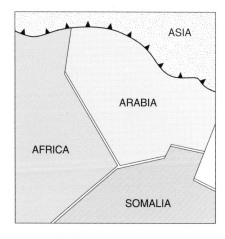

**Figure 4.27** Diagrammatic representation of the present-day configuration of plate boundaries defining the Arabian Plate. In Paleogene times, the constructive margins between the Arabian, African and Somalian Plates would not have existed and the three would have been merged into one large plate.

itself. The plate is currently rotating anticlockwise and in the vicinity of the Red Sea this rotation produces motion to the north-west. The absolute motion of the Arabian Plate, by contrast, is almost due north, driven by the last remnants of slab-pull beneath the Zagros mountains. It is also much greater than the motion of the African Plate (26 mm $yr^{-1}$ as opposed to 11–14 mm $yr^{-1}$), and this difference in absolute velocities and direction generates extension between the two plates.

This analysis of plate motions reveals that the slab-pull beneath the destructive plate margin located along the boundary between the Afro-Arabian Plate and the Eurasian Plate has been dominant throughout the past 80 million years or so. But even with this driving force the internal strength of the continental lithosphere is such that there must have been an additional factor for the Afro-Arabian Plate to split when and where it did.

● What do you think that extra factor was?

● The emplacement of the Afar mantle plume.

Our knowledge of the timing and composition of the Ethiopian flood basalts links them unequivocally to the underlying presence of a mantle plume. As with the present-day situation in the Kenya Rift and the East African Plateau, the emplacement of the Afar mantle plume generated excess relief and weakened the lithosphere by the emplacement of mafic magmas. Once extension had begun, the lithosphere became progressively weaker and extension was focused on a narrow zone of rift basins, which eventually evolved into the Red Sea. In the cases of the Kenya and Ethiopian Rifts, there are no destructive plate margins to drive further extension into true drift and so unless the plate configuration around the African Plate changes dramatically, it is unlikely to develop further, even to the stage of the Red Sea. In the case of the Red Sea, the effect of slab-pull continued after initial extension, allowing the Red Sea and Gulf of Aden to develop into the ocean basins we see today.

> **Question 4.22** What do you think might happen to the Arabian Plate and the Red Sea in the future?

## 4.6.7 Summary of Section 4.6

* The Red Sea region can be divided into three major plates: the Arabian, African and Somali Plates.

* These are separated by constructive margins of the African Rift, with an RRR junction located in the Afar Depression.

* Differential extension across the Gulf of Suez and the Red Sea basin is accommodated by the Dead Sea transform.

* The Arabian Plate is moving counter-clockwise (north-east) about a pole of rotation in the southern Mediterranean.

* Hot spot trails in the southern Atlantic show that the united Afro-Arabian Plate was moving counter-clockwise (NE) at about 30 mm $yr^{-1}$ from 80 to 30 Ma.

* The African Plate slowed significantly between 30 and 20 Ma, at the same time as the Red Sea first rifted.

* This slowing can be related to the isolation of the African Plate from the slab-pull of subduction beneath the Zagros mountains.

* Rifting of the Red Sea may have been partly triggered by the initial emplacement of the Afar mantle plume.

# Objectives for Section 4

Now that you have completed this Section, you should be able to:

4.1   Understand the meaning of all the terms printed in **bold**.

4.2   Describe the geological evolution of the Red Sea and surrounding region.

4.3   Outline differences in the composition, timing and eruption rate of basalt provinces relative to their tectonic environment.

4.4   Determine the age and duration of a sequence of flood basalts, using $^{40}Ar/^{39}Ar$ an d paleomagnetic data.

4.5   Outline the present plate tectonic boundaries relevant to the development of the Red Sea.

4.6   Use geological and fission track evidence to determine the age of lithospheric extension.

4.7   Evaluate the various influences on plate motion and continental break-up in the evolution of the Red Sea.

# 5  Flood basalts and continental break-up

## 5.1  Introduction

In the previous two Sections, you have followed the development of continental break-up through the rifting and initial drifting stages. In the Kenya Rift and the Red Sea, these processes are associated with basaltic magmatism related to lithospheric thinning (extension) over a mantle plume, which provides both heat that weakens the lithosphere, and uplift that increases local gravitational potential energy. The narrative that has developed through these two case studies implies that mantle plumes play a central role in continental break-up. But have mantle plumes been the prime cause of continental break-up globally and through geological time? If this can be demonstrated, then mantle plumes are largely responsible for the present disposition of the continents and ocean basins, defining where continents break to form new oceans. To investigate further the link between continental break-up and mantle plumes, we need to look back in time to magmatic provinces that now span intervening oceans and include basaltic magmas generated in both continental and oceanic regimes.

## 5.2  Continental flood basalts

The early development of the Red Sea was accompanied by the eruption of the Ethiopian flood basalts. They were erupted rapidly, were derived at least in part from a mantle plume and were eventually followed by separation of the African and Arabian Plates. The plume responsible for these basalts is still active beneath the region and its presence is evident in geophysical and topographic anomalies. But what if break-up had happened 100 Ma ago and the two continents had drifted 2000 km apart rather than a few hundred kilometres? What evidence would there be that a mantle plume had been active? Part of the answer is that only the basalts would remain; the topographic and geophysical effects (e.g. negative Bouguer gravity anomaly and high heat flow) would have disappeared. The only lasting evidence on land for the original presence of the plume at the site of continental break-up would be a thick sequence of flood basalts.

◉  What evidence would you expect to find on the ocean floor?

◉  The flood basalts and the site of the still-active plume might be linked by an aseismic ridge.

You have already encountered an example of this in your investigation of the motion of the Afro-Arabian Plate in Activity 4.2. But are there continental flood basalts of different ages that can be related to mantle plumes and continental break-up elsewhere? Figure 5.1 is a map of the world showing the locations of some continental flood basalt (CFB) provinces, and how some are linked to modern-day mantle plumes. They include the Afar plume, and the Ethiopian basalts which form the most recent of major CFB provinces associated with continental break-up. Another example is the Etendeka CFB province of southern Africa and their larger partner in South America, the Parana basalts. These are both linked to the island of Tristan da Cunha by aseismic ridges, the Rio Grande Rise to the west and Walvis Ridge to the east. Tristan da Cunha is an oceanic island underlain by a mantle plume and the aseismic ridges were generated when the Tristan plume lay beneath the Mid-Atlantic Ridge in a manner similar to that seen beneath Iceland today. Likewise, the flood basalts of eastern Greenland and the Tertiary volcanic rocks of north-western Britain can also be linked to the Iceland mantle plume. There are certainly flood basalts in different parts of the world and in many cases they can be linked to currently active

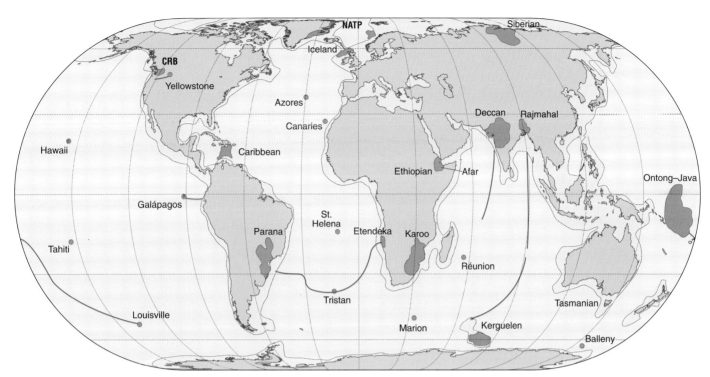

**Figure 5.1** Global distribution of continental flood basalts, some currently active mantle plumes and linking aseismic ridges. The present locations of mantle plumes are denoted by small reddish dots. NATP: North Atlantic Tertiary Province. CRB: Columbia River Basalts.

mantle plumes, but their relationship to continental break-up can only be determined by precise geochronology of both the basalts and the timing of continental rifting and drifting.

As you have seen in the case study of the Ethiopian flood basalts, radiometric dating can be applied effectively to dating flood basalts. However, it is more difficult to determine the age of continental rifting and separation.

As you will recall from the previous Section, fission track dating can be used to determine the age of rift flank uplift and this might be used to date rifting. However, rift escarpments are ephemeral features (geologically speaking), and are subject to erosion that eventually removes any major topographic expression. As a consequence, the interpretation of fission track ages becomes complex and more ambiguous as time progresses. Fission tracks provide a more precise age for uplift of a Tertiary event, such as the Red Sea flanks, than of an older event such as rifting associated with the opening of the South Atlantic. Geological methods of dating, such as offset of well-dated horizons by faults, can also be used, but it is often difficult to date the onset of rifting with precision.

By contrast, the onset of drifting is recorded in a more permanent and less ambiguous way by the age of the oldest ocean floor around the continental shelf. This age can be determined either directly by radiometric methods if samples are available or, alternatively, by using paleomagnetism. You should recall that the ocean floor records the orientation of the Earth's magnetic field when it is generated at a mid-ocean ridge and that the timescales of these reversals have been determined with a high degree of accuracy. You have already used this timescale in your analysis of the duration and date of the Ethiopian flood basalt (Figure 4.9). The geomagnetic timescale has been calibrated back to the Mesozoic and can be used to date areas of sea-floor without recourse to radiometric techniques.

Armed with these two geochronological techniques — radiometric dating and the geomagnetic timescale — you are now ready to examine in more detail the sequence of events that led to the break-up and dispersal of Gondwana, the supercontinent that dominated the Southern Hemisphere in the later Paleozoic and early Mesozoic. The various stages of this process were extensively reviewed by Bryan Storey (a British geologist who works in Antarctica) in the journal *Nature*, in 1995.

Now attempt Activity 5.1, in which you are asked to read this review and abstract critical information concerning the timing of basaltic volcanism, rifting and sea-floor spreading during the Jurassic and Cretaceous Periods.

### Activity 5.1  Mantle plumes, flood basalts and the break-up of Gondwana

Activity 5.1 involves reading a review paper published in the journal *Nature* on the break-up of Gondwana. You should allow about 2.5 hours for this Activity.

The results from Activity 5.1 show that there are numerous relations between mantle plumes, sea-floor spreading and continental break-up. Sometimes there are long delays between flood basalt episodes and drifting, at other times the gap is small. Sometimes flood basalts have a direct connection with active mantle plumes via aseismic ridges, sometimes not, and sometimes continental break-up does not involve mantle plume activity at all. These observations suggest that the process of continental break-up is variable, and to illustrate this variability the following Sections highlight aspects of two examples from Gondwana that show how mantle plumes can interact in varying ways with the overlying continental lithosphere with different consequences.

## 5.3  The Deccan flood basalts

It is probably easiest to start with the most recent episode of plume-related magmatism associated with Gondwana break-up, i.e. the eruption of the Deccan flood basalts in India. These are clearly related in both space and time to rapid motion of the Indian Plate over the Réunion hot spot, and the separation of the Seychelles microcontinent. They represent a good example of how mantle plumes can be involved in flood basalt generation during continental rifting. The Deccan flood basalts were emplaced about 65 Ma ago, close to the Cretaceous–Tertiary boundary, over a very restricted time period of about 1 million years.

> **Question 5.1**  The volume of the Deccan flood basalts is estimated to about 2 000 000 km³. (a) Calculate the mean eruption rate of the Deccan basalts. (b) How does this estimate compare with those from modern examples of plume-related basaltic magmatism (see Table 4.1)?

Very high eruption rates associated with the Deccan, and their coincidence with major faunal extinctions at the end of the Cretaceous (e.g. the dinosaurs), have led some workers to suggest a causative link between the two. Volatiles, especially $CO_2$ and $SO_2$, released into the atmosphere in large volumes from erupting basalts could have a long-lasting global environmental impact with major effects on food chains and ecosystems. But why were Deccan eruption rates so much higher than any plume-related volcanism today? How can they be explained within the context of global geodynamics?

Clearly, a mantle plume is involved. We have seen from Activity 5.1 that the Deccan basalts can be related via an aseismic ridge to the Réunion hot spot. However, the eruption rate on Réunion today is only about 0.1 km³ yr⁻¹, one to two orders of magnitude lower than that for the Deccan basalts. How can it be that a mantle plume currently feeding Réunion with magma at a modest rate could have given rise to the Deccan Province? This is particularly puzzling given that Réunion is located on relatively thin 60 Ma old oceanic lithosphere whereas the Deccan continental flood basalts are located on old and thick (>120 km) Archean continental lithosphere.

**Question 5.2**  Suggest factors that might contribute to the very high eruption rates of the Deccan basalts.

We know that the Deccan basalts are closely associated both in space and time with the rifting and drifting of the Seychelles Bank, but the question is whether infinite stretching over the Réunion plume was enough to generate the eruption rate in the Deccan province? You have already investigated one region where this is happening today — Iceland. We know from that case study that even with the Iceland plume beneath the Mid-Atlantic Ridge, the eruption rate is still only 0.24 km³ yr⁻¹ which is much less than even the lowest estimate of basalt production in the Deccan. It is therefore unlikely that the Deccan basalts resulted simply from extension, or even sea-floor spreading, over a mantle plume.

What about the alternative, that the potential temperature of the Réunion plume was greater in the past? A currently popular suggestion is that flood basalts are related to the impact of mantle plumes. All modern-day examples of plume-related magmatism, be they oceanic as on Hawaii, Iceland, Tristan da Cunha etc., or continental as in the African Rift, are the surface manifestation of so-called 'steady-state' mantle plumes — in other words, plumes that have a long history of volcanism produced at a more or less constant rate. For example, eruption rates along the Hawaiian chain have remained between 0.1 and 0.2 km³ yr⁻¹ throughout the past 10 Ma. However, mantle plumes do not last forever — they have to begin at some time and they eventually die; and, while the complete history of a mantle plume has not been recorded in detail, the progression of ages along aseismic ridges implies that they may last for 100 Ma or more (e.g. the Tristan plume associated with the Parana–Etendeka flood basalts).

How then does a mantle plume begin? Laboratory experiments using suitable analogue materials suggest that a starting plume initiates as a thermal instability at a boundary layer deep within the mantle. With time, hot material at the boundary layer begins to upwell and rise through the mantle. This initial surge of material develops into a large spherical head that rises through the mantle, followed by a much narrower tail (Figure 5.2). The plume head is initially a few hundred kilometres across but expands by entrainment of surrounding mantle

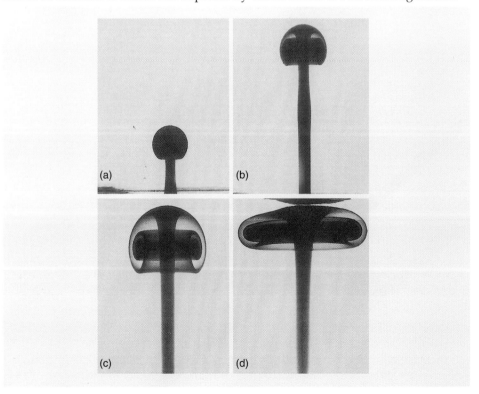

**Figure 5.2**  A laboratory analogue of the initiation of a mantle plume. The original laboratory experiment involved heating the basal layers of a tank of two viscous fluids. The lower layer is slightly denser than the upper layer but when heated it expands and its density decreases. Eventually, its density drops below that of the upper layer and a large mass of material starts to rise to the top of the tank.

material and eventually flattens out to form a disc of hot material below the lithosphere. The overall profile is like a growing mushroom and the dynamics are similar to those that operate in the atmosphere during the eruption of large volcanic plumes or the mushroom clouds of nuclear explosions.

Within the Earth, thermally driven upwellings such as mantle plumes are initiated at a boundary, which may be the core–mantle boundary or the boundary between the upper and lower mantle. In Figure 5.2, the initial 'plume head' is followed by a 'plume tail', a narrow jet of hot material rising from the lower layer. It has been suggested that flood basalt provinces result from the large volume plume head while oceanic islands and aseismic ridges are the product of the more continuous but less voluminous plume tail.

You have seen in the case study of the Kenya Rift how a mantle plume produces dynamic uplift of the overlying lithosphere and how this uplift contributes to extension. The emplacement of a large plume head beneath the lithosphere will also produce uplift on a large scale and it has been suggested by some that this effect in itself could cause lithospheric extension. Whether uplift alone is enough to cause rifting without the additional effects of far-field plate tectonic or intraplate forces remains a matter for debate. What is clear is that extension will lead to decompression of the underlying plume head material which will then melt. Because the plume head is both large and hot, there is great potential in this system to produce large volumes of basaltic material very rapidly. This combination of extension over a starting plume head is the most likely explanation for the rapid eruption of the Deccan flood basalts.

⬤ Can you think of any tests for this model?

⬤ The basalts should show evidence of being derived from a mantle plume source.

You should recall from the Ethiopian flood basalts that the plume-related HT2 basalts had smooth mantle-normalized trace element abundance patterns, very similar to those of many ocean island basalts. The same is true for the Deccan basalts as shown in Figure 5.3, where mantle-normalized trace element abundances of the Cretaceous Deccan are compared with recent basalts from Réunion. As you can see, both patterns are comparable but there are subtle differences.

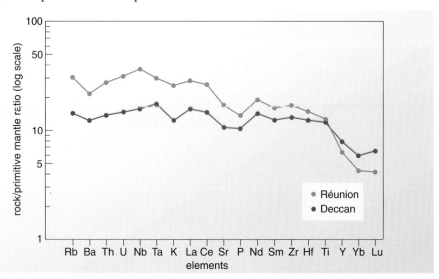

Figure 5.3   Mantle-normalized incompatible element diagram comparing Deccan basalts with basalts from Réunion.

Question 5.3   Study Figure 5.3. (a) What are the two most important differences between the trace element profiles of the Deccan basalt and the basalt from Réunion? (b) Assuming both basalts are primary or near-primary composites and both were derived from the same source region, what are the causes of these differences in trace element abundances?

The lower melt fractions implied for the Réunion basalts are consistent with their generation beneath oceanic lithosphere, restricting melting to depths where garnet is a residual mineral. In the case of the Deccan basalts, melting continued to much shallower depths because of the effects of lithospheric thinning during the rifting and drifting of the Seychelles Bank. Hence, the influence of garnet on melts generated at greater depths is overwhelmed by the much greater melt fractions derived from shallower depths.

Finally, it is useful to explore the differences between the relative abundances of different incompatible elements in the Deccan and other oceanic basalts. You should recall from Section 2 that while the melt fraction controls the concentration of an incompatible element in the melt, it has little effect on incompatible element ratios as long as the melt fraction is large. Hence, on a graph of La/Nb against La/Ba (Figure 5.4), MORB occupies a relatively restricted field because La, Nb and Ba are all highly incompatible elements and the source region of MORB is globally homogeneous. Because all three elements are highly incompatible in the mantle, the ratios plotted are not fractionated during mantle melting and so values of the basalts are very similar to those of their mantle source regions. Oceanic basalts define a positive trend on this diagram, reflecting the systematic fractionation of Ba and Nb from La by melting within the mantle. MORB are derived from a depleted source region (Section 2) and so plot at low values of La/Nb and La/Ba, whereas OIB are derived from more enriched source regions and plot at higher values of these two ratios. This slight difference in the position of MORB and OIB reflect differences in their source compositions. Adding the Deccan and Réunion data shows that both overlap with the OIB field, confirming that both are derived from the same source region that was similar to the source regions of other oceanic islands.

Thus, the evidence from Réunion and the Deccan emphasizes the close link between the two provinces and in general supports the idea that the Deccan basalts represent magmas generated when the Réunion plume made its initial contact with the lithosphere 65 Ma ago.

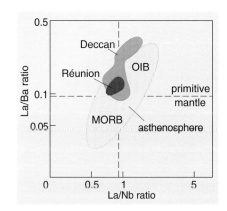

**Figure 5.4** La/Nb and La/Ba ratios in oceanic basalts, Réunion and the Deccan.

## 5.4   The Parana–Etendeka flood basalts

The Parana–Etendeka basalts are closely linked to a mantle plume currently located beneath Tristan da Cunha by an aseismic ridge (Figure 5.1). However, unlike the Deccan basalts, they pre-date sea-floor spreading in the South Atlantic by up to 10 Ma, and they also have a more protracted eruption history. They have a volume of about 2 000 000 km$^3$, but were erupted over a period of about 10 Ma between 138 and 128 Ma, giving a mean eruption rate of ~0.2 km$^3$ yr$^{-1}$ and a peak eruption rate of 0.4 km$^3$ yr$^{-1}$. Yet, because they pre-date sea-floor spreading and are underlain by Proterozoic lithosphere, there is a problem in explaining why any magma was produced at all. Referring back to Figure 3.27, with no extension (i.e. a $\beta$ factor of 1), and a lithospheric thickness of 130 km or more, a mantle plume needs to have an unrealistically high potential temperature before it will melt. But did the plume melt?

As with the Deccan basalts, this question can be partly answered by reference to incompatible element abundances and comparison with ocean island basalts. The average trace element contents of typical Parana lavas are illustrated in Figure 5.5 and compared with an average of modern-day basalts from Tristan da Cunha.

- Are the Parana lavas similar to or different from the Tristan basalts?

- They are different. The Parana basalts show depletions of Nb, Ta and Ti relative to the light REE Ba and Th.

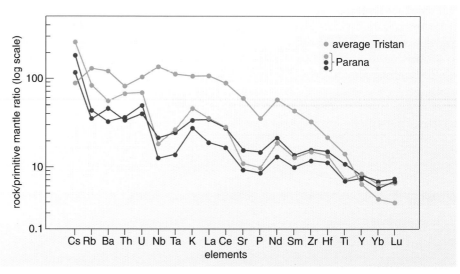

**Figure 5.5** Mantle-normalized incompatible element diagram comparing trace element contents of three different Parana basalt magma groups with basalts from Tristan da Cunha.

These differences in trace element composition can be further illustrated using the plot of La/Ba against La/Nb. As has already been emphasized, the Deccan basalts plot close to the trend defined by oceanic basalts, consistent with their derivation from a mantle plume source region. By contrast, the Parana basalts fall off this trend, having high La/Nb ratios but low La/Ba (Figure 5.6), and the conclusion is that they were derived from a source region that is distinct from mantle plumes.

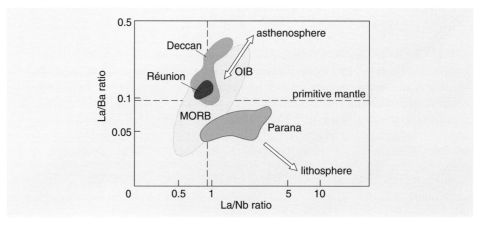

**Figure 5.6** A plot of the La/Nb and La/Ba ratios in oceanic and selected continental flood basalts.

Further insight into the nature of this source region can be gained from the major elements. You will recall from Section 2 that major element compositions of primary basalts depend on the depth (pressure) and temperature of melting

● What is the effect of increasing the pressure of the melting regime on the $SiO_2$ content of a basaltic melt?

● Higher pressure leads to lower $SiO_2$.

Although the majority of the Parana flood basalts are evolved, with MgO contents < 8 wt % and Mg# < 60, estimates of their primary compositions suggest $SiO_2$ contents of 47–49%.

> **Question 5.4**  (a) Based on the relationship between $SiO_2$ and pressure of melt generation (Figure 2.17), what pressures and depths of melting do $SiO_2$ contents between 47–49% imply? (b) Are these depths within the mantle or crust? (c) What is the mineralogy of the mantle at these depths? (d) If there has been no lithospheric thinning, do these depths correspond to the lithosphere or asthenosphere?

The major element compositions therefore point to a shallow melt regime, as with the LT basalts in the Ethiopian flood basalts. Yet you should recall that the flood basalts pre-date sea-floor spreading by about 10 Ma and so the lithosphere could not have thinned enough to allow the mantle plume to melt. This is consistent with the trace element data which also imply that the plume was not melting. These observations have led to some suggestions that rather than the plume melting, it is the mantle lithosphere that melted. However, in most models of mantle melting you have encountered so far, the lithosphere is that part of the mantle in which heat is transferred by conduction and is always below the peridotite solidus. How, then, can it melt?

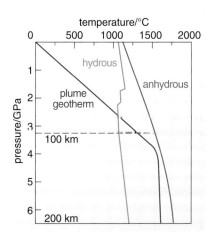

**Figure 5.7** Hydrous and anhydrous peridotite solidi, compared with a mantle plume geotherm ($T_p = 1500\,°C$).

The solution to this problem lies in the mineralogy of the continental mantle lithosphere and the effects of volatile compounds such as water and carbon dioxide on the mantle solidus. The mantle solidus with which you are by now familiar is based on an anhydrous mineral assemblage of olivine, pyroxenes and an aluminous phase. However, xenoliths of mantle material brought to the surface in deep-sourced continental volcanic rocks (e.g. kimberlites and alkaline mafic rocks) often contain crystals of hydrous minerals such as amphibole and mica. Although these minerals are generally present in low modal abundances (<5%), they have a dramatic effect on the solidus temperature of the mantle. This is illustrated in Figure 5.7 which compares the anhydrous mantle solidus with that of amphibole and mica-bearing peridotite. Quite clearly, mantle with volatile-bearing phases melts at a much lower temperature than anhydrous mantle.

For this hydrous mantle to melt, it must be heated and, in the case of the Parana, the heat is supplied by conduction from the underlying Tristan mantle plume. However, rocks are poor conductors of heat and so melting only begins after a period of heating, estimated to be about 10 Ma. The mantle plume doesn't melt because the lithosphere is thick enough to prevent it from rising to the depth necessary for decompression melting. Melting in the lithosphere then continues for as long as there is mantle lithosphere available to melt while the melting rate is related to the rate at which heat is transported into the lithosphere by conduction and local advection by rising magma. Magmatism is therefore prolonged, in contrast to the rapid generation and eruption caused by the plume impact model. In addition, conductive heat transfer into the mantle lithosphere reduces its thickness in a process described as thermal erosion. This is an alternative to tectonic thinning and you have already encountered it in the example of the Kenya Rift. Uplift consequent on plume emplacement induces rifting and eventually drifting takes place if the overall plate tectonic configuration is right.

Hence, many of the important features of the Parana flood basalt province such as its long duration, modest eruption rates, lithospheric source regions and relationship to the onset of sea-floor spreading, can be explained in a model in which the mantle plume acts as a source of heat beneath thick continental lithosphere.

## 5.5 Concluding comments

In this Section, we established a strong link between the eruption of continental flood basalts and continental break-up. However, before we leave this subject, it is important to recognize that there are some notable exceptions to the general rule implied by this association. One of these is the largest flood basalt province of all — the Siberian Traps (Figure 5.1). These basalts are of Permo-Triassic age and are located, as their name suggests, in Siberia. They cover a vast area and have a total volume greatly in excess of $2\,000\,000\,km^3$. Magnetostratigraphy shows evidence for only a few reversals, consistent once again with $^{40}Ar/^{39}Ar$

ages that suggest a total duration of less than 2 Ma. Yet despite these typical CFB characteristics, they were not associated with continental break-up. The Siberian traps were emplaced within a large sedimentary basin with up to 5 km of Paleozoic sediments and are covered with a similar thickness of Mesozoic and Cenozoic rocks. Lithospheric extension has produced this basin but sea-floor spreading was never initiated. Clearly, a large mantle plume of itself does not necessarily split a continent.

At the opposite extreme is the rifting and drifting of Australia from Antarctica. This episode in the break-up of Gondwana is described in the review article (Activity 5.1) and was not accompanied by any magmatism until sea-floor spreading generated true oceanic crust between the separating continental blocks. There is no evidence at all for the involvement of a mantle plume. The driving force for this event is probably related to far-field intraplate or plate tectonic forces (e.g. slab-pull) and is probably the best example there is of a passive rift developing into continental break-up.

In between these two examples, there is a wide spectrum of cases, some of which you have now investigated, in which flood basalts and aseismic ridges testify to an important but not exclusive association between plumes and continental break-up. As with much of Earth sciences, the plethora of variable factors that can be introduced in any specific case makes each one individual, although the search for a unifying model continues.

## 5.6   Summary of Section 5

- Flood basalts show a close spatial and temporal association with many phases of the break-up of Gondwana.

- As break-up proceeds, the time gap between flood basalt magmatism and sea-floor spreading decreases.

- Different flood basalt provinces have different characteristics, especially their duration, eruption rate and mantle source regions.

- Early flood basalts such as the Parana are dominated by lithospheric mantle sources, whereas later provinces (e.g. the Deccan) show more involvement of mantle plume sources.

- The lack of continental break-up associated with one of the largest flood basalt provinces (Siberia) implies that mantle plumes cannot alone lead to break-up and sea-floor spreading.

- The lack of evidence for mantle plume activity in the rifting of Australia from Antarctica implies that mantle plumes are not essential for continental break-up.

## Objectives for Section 5

Now that you have completed this Section, you should be able to:

5.1   Understand the meaning of all the terms printed in **bold**.

5.2   Describe and explain the relative position of flood basalts, aseismic ridges and mantle plumes in relation to the break-up of Gondwana.

5.3   Outline the variable relationships between continental break-up and the timing, duration and composition of flood basalts.

5.4   Discuss the evidence for models of basalt generation in different flood basalt provinces.

# Answers and comments to questions

**Question 2.1** The porphyritic texture suggests a period of slow cooling that allows the growth of large crystals, followed by a period of rapid cooling to produce the fine-grained groundmass. These two periods might relate to initial cooling in a subsurface magma chamber followed by cooling after eruption. The intersertal texture suggests a period of steady cooling, possibly in the centre of a large flow after eruption, or in a dyke or sill.

**Question 2.2** (a) Basalt 1 has 47.23% $SiO_2$ and total alkalis of 4.82% (2.91 + 1.91), and is clearly an alkali basalt. Basalt 2 has 48.88% $SiO_2$ and 2.61% total alkalis, and plots in the subalkaline basalt field. Basalt 3 has 51.08% $SiO_2$ and 3.46% total alkalis, and plots below the A–B dividing line on Figure 2.3 and so is a subalkaline basalt.

(b) Basalt 1 has normative nepheline (and no quartz or hypersthene), and is therefore an alkali basalt. Basalt 2 has normative hypersthene but has no quartz or nepheline, and is therefore an olivine tholeiite. Basalt 3 has normative quartz (and normative hypersthene), but no nepheline, and is therefore a quartz tholeiite.

(c) Basalt 1 is classified as an alkali basalt in the alkali–silica plot and also as an alkali basalt by its normative mineralogy. Basalts 2 and 3 are both classified as tholeiitic basalts (tholeiites) on the alkali–silica plot, and they are also both classified as tholeiites using their normative mineralogy — although subdivided into an olivine tholeiite (basalt 2) and a quartz tholeiite (basalt 3) by their degree of silica saturation. It can be concluded that both classification schemes agree with each other when classifying basalts as either alkali basalts or tholeiitic basalts.

**Question 2.3** Figure 2.4 suggests that alkali basalts are found on oceanic islands and subalkaline rocks at mid-ocean ridges and island arcs. However, Figure 2.3 reveals the presence of both subalkaline and alkali basalts on Hawaii, an oceanic island. In general, basalts from mid-ocean ridges are tholeiitic and rarely alkaline whereas both tholeiites and alkali basalts are found on oceanic islands. In general, large oceanic islands such as Iceland and Hawaii contain both basalt types whereas small oceanic islands such as the Azores and the Cape Verde islands in the Atlantic are dominated by alkaline basalts. Basalts from island arcs are most often silica saturated and you will encounter these in Block 3.

**Question 2.4** (a) Rock names are shown on the alkali–silica plot with the grid (Figure 2.2), and these rocks (c. 56% $SiO_2$ and c. 10% total alkalis) plot within the trachyandesite field. (b) Using Figure 2.2, the island arc rock with the highest $SiO_2$ content (c. 60%) and ~5% $Na_2O + K_2O$ plots within the andesite field.

**Question 2.5** (a) Melt a will crystallize albite at about 1110 °C followed by albite plus nepheline at the Ab–Ne eutectic at a temperature of 1080 °C. Melt c also first crystallizes albite but this is followed by crystallization of albite plus quartz at the Ab–Q eutectic at ~1060 °C. If the crystals and melt become separated, then it is possible to produce liquids of very different compositions from starting compositions that are very close to one another. (b) None. Neither equilibrium nor fractional crystallization will produce a Ne-bearing residual liquid from one that starts on the Q side of the albite composition. (c) Liquid b lies on the albite composition. It will therefore crystallize only albite at its liquidus temperature which is 1120 °C.

**Question 2.6** The results of this question are analogous to those of the previous one. All compositions lie in the Di field and so will crystallize diopside first. However, liquid a will evolve away from the Di apex towards the Di–Ab cotectic where it will start to crystallize Ab together with Di. The liquid will then migrate down temperature towards the eutectic x at which point it will start to crystallize Ne together with Di and Ab. Similarly, composition c will evolve towards the Di–Ab cotectic but on the Q side of the Di–Ab join. Once on the cotectic, the liquid will crystallize Di + Ab and migrate towards eutectic point z, where Di and Ab will be joined by Q. Composition b lies on the Di–Ab join which is a binary eutectic and so crystallization of Ab + Di will continue at point y and the liquid will migrate no further. Thus, the three similar starting compositions can produce three very different final liquid compositions, one with Ne, one with Q and one with neither Ne nor Q.

**Question 2.7** All mantle minerals are solid solutions between MgO and FeO end members, e.g. olivine Fo–Fa, clino- and orthopyroxene En–Fs. The system illustrated contains no iron. Secondly, there is no aluminium in this system and we have noted above that most peridotites contain an Al-rich mineral, albeit in often small proportions. Thirdly, the system contains no titanium, sodium or potassium.

**Question 2.8** Pathway a–b–c represents mantle that is rising along the convective geotherm (point a), and then rises adiabatically to cross the solidus at point b. Above point b, the mantle will partially melt and if the upwelling mantle (now solid + melt) continues to rise adiabatically it will reach the surface at point c. Pathway x–y–z represents hotter mantle that is rising from depth, and therefore it does not follow the convective geotherm (i.e. it is displaced towards higher temperatures). As a consequence of its higher temperature, this hotter mantle crosses the solidus at greater depths (pressures) at point y, and if the upwelling mantle (now solid + melt) continues to rise adiabatically it will reach the surface at point z, which is at a higher temperature than point c. Pathway a–b–c represents typical melting beneath oceanic spreading ridges, with melting taking place at relatively shallow depths, whereas pathway x–y–z represents an upwelling mantle plume with melting taking place at relatively higher pressures (and temperatures).

**Question 2.9** (a) Mantle rising adiabatically from point C will intersect the solidus at greater pressure (i.e. greater depth) than mantle B. (b) The degree of partial melting increases the higher the temperature above the solidus. Therefore, at all pressures where melt is present, mantle C produces a greater degree of partial melt. (c) Mantle C, with its higher potential temperature, will produce greater proportions of melt over a wider range of pressures than mantle B. Therefore, the total volume of melt produced by mantle at the higher potential temperature (C) will be greater than that produced by mantle at lower potential temperature (B).

**Question 2.10** (a) From Figure 2.16 the mantle potential temperature that would generate a 7 km thick crust is 1280 °C. (b) A mantle potential temperature of 1280 °C approximates to a maximum melt fraction (using Figure 2.16b) of 0.25 (equal to 25% by mass). (c) The mantle potential temperature of 1280 °C equates to a maximum depth of partial melting at c. 42 km. This implies that melting would *begin* at this depth within the spinel–lherzolite stability zone of Figure 2.10.

**Question 2.11** This is tackled by establishing the points on the solidus immediately above both $Fo_{88}$ and $Fo_{92}$. These represent the compositions of the (solid) olivine crystals. The co-existing liquid compositions are determined by drawing horizontal tie-lines from each of these points to the liquidus. And from the two points on the liquidus, read off from the scale on the x-axis the wt % of Fo, which should be $Fo_{63}$ to $Fo_{73}$.

**Question 2.12** None of the basalts can be regarded as a primary magma as all have Mg# vales <65, although basalt 2 with a value of 64.3 is close to being considered primary. Actual values of Mg# are: basalt 1, 51.1; basalt 2, 64.3; and basalt 3, 46.5.

**Question 2.13** The experiments only explore pressures up to 3 GPa whereas melting may take place at higher pressures. At the pressures covered, garnet is not a stable mineral phase and so the effects of this mineral on the melt composition are unknown. Fractional crystallization of a small amount of olivine will reduce the MgO content of a primary melt very quickly.

**Question 2.14** (a) Table 2.3 shows that basalt 2 contains 48.88% $SiO_2$ and 7.37% MgO. Using Figure 2.17a ($SiO_2$ versus pressure) gives a pressure of about 2.0 GPa, equivalent to a depth of about 60 km, and using Figure 2.17b (MgO versus temperature) gives a temperature of about 1250 °C. (b) Using the pressure–temperature diagram (Figure 2.10), this range of temperatures and pressures probably lies within the spinel–lherzolite stability field (just within the sub-solidus zone).

**Question 2.15** (a) The $SiO_2$ content of 46.4% implies a pressure of 2.5–3.0 GPa and hence a depth of 75–90 km while the MgO content of 9.47% suggests a temperature of about 1300 °C. (b) Basalt B gives a lower pressure (2.0–2.5 GPa, 60–75 km) but a higher temperature (1400 °C).

**Question 2.16** (a) If $C_0$ increases, $K_2O$ also increases because $C_0/C_1$ = constant = D. (b) If D increases, $K_2O$ decreases because more of the trace element is held back in the source region. (c) If F increases, $K_2O$ decreases because Cl is inversely proportional to F.

**Question 2.17** Higher $K_2O$ in alkali basalts relative to tholeiites suggests either that the source of alkali basalts is enriched in potassium, or the D-value for K is lower during the generation of alkali basalts or the melt fraction is lower.

**Question 3.1** Both the Tertiary rift systems and Cenozoic volcanic activity occur primarily in the non-cratonic regions. Where rifts intersect the cratons, they rapidly die out and the craton remains relatively unaffected.

**Question 3.2** We know from the oceans that extension leads to decompression melting of the underlying mantle and the same could be occurring beneath continental rifts. Furthermore, some of the rift faults penetrate deep within the Earth, allowing easy passage of any magmas produced at depth to the surface.

**Question 3.3** Cratons are old regions of the crust that have been tectonically inactive for 2.6 Ga. They are characterized by low geothermal gradients and are consequently much stronger than the younger, hotter lithosphere of the mobile belts. In addition, they have much thicker lithosphere, extending to possibly 200 km depth. The greater strength of the cratons means that they will not stretch as easily as the mobile belts during extension while their greater thickness suppresses melting in the underlying asthenosphere.

**Question 3.4** There is a broad tendency for nephelinites and carbonatites to occur on the craton or remobilized craton margins. Elsewhere, volcanism is dominated by basalt–trachyte–phonolite associations, although nephelinitic activity is currently present in the far north (Turkana).

**Question 3.5** There is a remarkable similarity between the gravity map and topography but no clear connection between gravity and basement geology. Across the whole of Africa, areas of high elevation are characterized by negative Bouguer gravity anomalies and two of the strongest anomalies occur over the Ethiopian and East African plateaux.

**Question 3.6** Crustal rocks tend to melt at temperatures below 1000 °C while mantle rocks, dominated by olivine, lose their strength. Thus, the lithosphere will become weaker.

**Question 3.7** Natural seismicity in the Ethiopian and Kenya Rifts is surprisingly limited. However, there are more earthquakes in northern Tanzania and the highest earthquake density occurs along the Western Rift. Focal depths are also shallow in Kenya (down to 12 km) and deeper in northern Tanzania (20–30 km) and along the Western Rift (20–40 km). These are both areas where the rift either cuts across the cratonic crust or an area of more limited volcanic activity. The intensity and depth of seismic events in the cratonic regions reflects the depth to which the crust behaves in a brittle fashion, because it is cooler. The higher geothermal gradients beneath the Kenya Rift mean that the crust behaves in a brittle fashion only at relatively shallow depths.

**Question 3.8** Summary of geophysical knowledge prior to 1985:

- Broad negative Bouguer anomaly across the Kenya Dome and the whole of the East African Plateau, suggesting low density material at depth, probably in the mantle.

- Small positive gravity anomalies associated with central volcanoes on the rift floor, suggesting high density rocks below them but not elsewhere in the rift.

- High heat flow associated with central volcanoes in rift floor, normal heat flow on rift flanks. Implying high temperatures at the Moho beneath the rift axis and a largely ductile mantle below — possibly no mantle lithosphere

- Earthquakes at shallow depths (<12 km) beneath rift axis associated with major faults, suggesting shallow depths of brittle–ductile transition in crust. Deeper earthquakes in northern Tanzania (and in the Western Rift) reflect the stronger, more brittle cratonic lithosphere.

**Question 3.9** The decrease in the amount of extension southwards reflects the general sense of southward propagation of the rift. It implies that the earlier rifting starts the more extension has gone on. It might be considered analogous to tearing a piece of paper from one side to another.

**Question 3.10** The fast seismic velocities at all depths beneath the craton imply cold material to depths of 160 km and perhaps more. Likewise, cold mantle extends to layer 5 beneath the mobile belt (95–125 km) below which velocities are slower and mantle temperature consequently higher. These seismically fast regions probably represent undisturbed mantle lithosphere. The slowest velocities are found in mantle directly beneath the rift and the boundaries between slow (rift) mantle and fast (lithospheric) mantle show that hot mantle is restricted to those regions beneath the rift.

**Question 3.11** The total variation in velocity is 12% and a maximum of 6% can be attributed to temperature variations in solid mantle. Therefore, a minimum of 6% velocity variation is due to the presence of partial melt. Reading from

Figure 3.18b, this implies the presence of 3–6% partial melt. However, if all of the 12% variation in seismic velocity is due to the presence of melt then the melt fraction increases to 8%.

**Question 3.12** (a) The present-day rift has tended to evolve towards a narrow, full graben geometry, as shown in Figure 3.24a. In addition, the mantle lithosphere has been thinned beneath the rift axis and replaced by hot mantle from below. The symmetrical nature of deformation throughout the whole lithosphere points to pure shear. (b) During the Neogene, extension was accommodated in asymmetric half-graben and so deformation by simple shear might be more appropriate.

**Question 3.13** At mid-ocean ridges, the lithosphere is very thin, hence large amounts of tholeiitic melt can form at shallow depths within the mantle. Beneath the Kenya Rift, the lithosphere is much thicker and so melting is restricted to greater depths and the melt fraction is lower. Hence melts are more alkali-rich.

**Question 3.14** (a) The bulk compositions vary from 36–50 wt % $SiO_2$ and 2–8 wt % $Na_2O + K_2O$. There is a general tendency for the nephelinites to have lower silica and higher alkali contents than the basanites whereas the basanites have similar alkali contents to the alkali basalts, but lower silica. (b) The boundaries do not correspond precisely with the normative classification. This lack of agreement is due to the way in which normative minerals are calculated. Norms depend on the amount of $Al_2O_3$ and $Fe_2O_3$, which are not part of the compositional classification shown in Figure 3.26, as well as the amount of $SiO_2$ and $Na_2O$ in a rock. This example serves to illustrate how different classification systems can lead to different names and how, for example a basanite defined under one system may be termed a basalt or a nephelinite under another.

**Question 3.15** $D = 0.005$, $C_0 = 0.5$.
For $F = 1\% = 0.01$, $C_l = 0.5/(0.005 + 0.01(1–0.005)) = 33.4$ ppm.
For $F = 5\% = 0.05$, $C_l = 0.5/(0.005 + 0.05(1–0.005)) = 9.1$ ppm.
For $F = 10\% = 0.1$, $C_l = 0.5/(0.005 + 0.1(1–0.005)) = 4.8$ ppm.
For $F = 25\% = 0.25$, $C_l = 0.5/(0.005 + 0.25(1–0.005)) = 2.0$ ppm.

**Question 3.16** Rearranging the partial melting equation gives the following expression for $F$:

$$F = (C_0/C_l–D)/(1–D)$$

Substituting the values given provides the following melt fractions:

nephelinites $F = (0.5/75–0.005)/(1–0.005) = 0.0017$, i.e. ~0.2%.
basanites $F = (0.5/45–0.005)/(1–0.005) = 0.0061$, i.e. ~0.6%.
basalts $F = (0.5/25–0.005)/(1–0.005) = 0.015$, i.e. ~1.5%.

**Question 3.17** The only mantle mineral that takes up Y and Yb is garnet with $K_d$-values of 2 and 4 for Y and Yb respectively. The phase relations for peridotite are illustrated in Figure 2.10. They show that garnet is stable only at high pressures, corresponding to depths greater than 75 to 90 km.

**Question 3.18** The average silica content of the basalts corresponds with experimental melts produced between 2.5 and 3.0 GPa pressure or 75–90 km depth. At these depths, the mantle has a mineralogy that contains both garnet and spinel, the latter being in the process of reacting with clinopyroxene to form garnet. The basanite has a silica content less than that observed in any of the experimentally derived melts. However, only a little extrapolation of the experimental data suggests their derivation from depths in excess of 90 km, well within the garnet stability field. Finally, the average nephelinite has an extremely low silica content of 42.7%, which is far beyond the limits of the experimental range covered. If the relationship does hold to greater pressures, then the low silica content of the nephelinites is entirely consistent with an origin from depths in excess of 100 km.

**Question 3.19** Yes. Decompression melting results from the upwelling of hot mantle into a lower pressure environment. As a parcel of mantle crosses the solidus, due to a drop in pressure, it begins to melt and the further it rises the more it melts, i.e. the greater the melt fraction. Thus shallow melts will reflect larger melt fractions than melts derived from greater depths. This is what we see in the Kenyan magmas. The upper limit of melting is controlled by the base of the lithosphere which beneath the rift is at about 80 km. However, some melts appear to have been derived from shallower depths than this, implying either greater thinning of the lithosphere beneath the rift or melting of the mantle lithosphere (or both). The bottom of the melting regime is at least at 125 km. Such depths of melt generation are only possible in mantle with an elevated potential temperature, i.e. a mantle plume. Finally, the thickness of the lithosphere and the observation that mantle melting occurs requires high mantle potential temperatures, regardless of the amount of melt generated.

**Question 4.1** Over much of the length of the Red Sea, the Arabian/African coastline 'fit' is remarkably good. This is emphasized in Figure A4.1, a reconstruction of the best coastline fit. The two land masses could have once been joined.

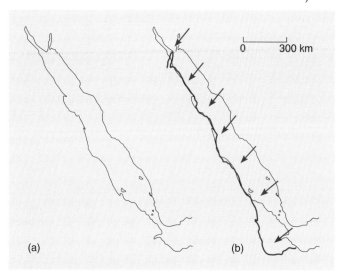

**Figure A4.1** Answer to Question 4.1. (a) Outline of the Red Sea. (b) Reconstructed coastline fit made by moving Arabia south-west towards Africa.

**Question 4.2** (a) Figure 4.1 shows a constructive margin extending to the western extremity of the Gulf of Aden but stopping there. There is another constructive margin along the axis of the Red Sea and also a zone of earthquakes along the East African Rift system. (b) For rifting to be possible in the Red Sea, Gulf of Aden and East African Rift, all must be areas of tensional stress. Figure A4.2 shows a first approximation of the stress patterns involved, with arrows drawn at right angles to the rift zones representing direction of stress and/or plate movement.

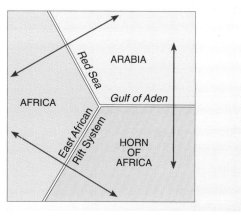

**Figure A4.2** Answer to Question 4.2. Possible constructive margins and stress patterns around the Red Sea.

**Question 4.3** (a) The best evidence concerning the age of the Red Sea basin is that Tertiary (Neogene) sediments occur along the flanking coastal plains and that these directly overlie Precambrian crystalline basement. This suggests, but does not prove, that the basin the Red Sea occupies is of Tertiary age (option D). (b) We can deduce from Figure 4.3 that the crust flanking the Red Sea was produced 500–1200 Ma ago and consists of granitic and metamorphic rocks overlain further away from the Red Sea by Phanerozoic sediments and Tertiary igneous rocks. So, option A is the best choice as it describes most of the crust.

**Question 4.4** No. Most of the Phanerozoic seems to have been remarkably quiet. Of the Paleozoic, there remain only isolated sedimentary remnants and any igneous rocks of Paleozoic age are not of volumetric significance.

**Question 4.5** The great thickness of evaporites is strong evidence for subsidence of the floor of the Red Sea basin while evaporites were being deposited.

**Question 4.6** Yes. Tholeiitic rocks occur in those areas of extreme extension where oceanic crust has formed (Red Sea axial trough, Gulf of Aden) or where the continental crust has thinned dramatically (Afar Depression). Transitional basalts are found in the remaining parts of the Afar Depression and the Ethiopian Rift where amounts of extension are more limited. Alkali basalts are found in the southern parts of the Ethiopian Rift and in the Saudi Arabian Peninsula where amounts of extension are very low (rift) or non-existent (Arabia). These variations in composition are consistent with the observed amounts of extension. The exception to this is in the Ethiopian Highlands, to the west of the rift, and the Yemeni plateau where transitional basalts occur but with no obvious link to extensional features.

**Question 4.7** (a) The oldest age is that of the lowest sample in the sequence — $30.8 \pm 0.4$ Ma and so the maximum possible age is 31.2 Ma. The youngest age, however, is one of the most precise, coming from a feldspar in the capping trachytic tuff. This is $29.7 \pm 0.1$ Ma, implying a minimum age of 29.6 Ma. The maximum duration is the difference between these, two i.e. 1.6 Ma. The minimum duration is given by the difference between 30.8–0.4 = 30.4 Ma and 29.7 + 0.1 Ma = 29.8 Ma, i.e. 30.4 – 29.8 = 0.6 Ma. (b) There are 43 flows in the section therefore the mean interval is 1.6/43 Ma = $c$. 40 000 years.

**Question 4.8** (a) Chron 11 includes the age 30.1 Ma, so this is probably the correct correlation. (b) The adjacent chrons must be chron11r and chron10r. The total duration of these chrons is from 30.8 Ma to 28.9 Ma, a total interval of 1.9 Ma. (c) Our estimate from Ar geochronology was 1.6 Ma. Therefore, the two estimates are very close.

**Question 4.9** (a) We estimated the volume of the continental flood basalts in Ethiopia and Yemen to be 750 000 km$^3$. If the duration is 1.6 Ma then the eruption rate is simply 750 000/1 600 000 km$^3$ yr$^{-1}$ or 0.47 km$^3$ yr$^{-1}$. (b) It is considerably larger than anything active today, being about three times the eruption rate of Hawaii and twice the eruption rate of Iceland.

**Question 4.10** The sedimentary record of the Red Sea basin suggests that it started to form possibly as long ago as 30 Ma. The Afar basalts have formed subsequent to that time (by decompression melting of the underlying Afar plume) and so can be removed in any continental reconstruction. The amount of extension across Afar is difficult to estimate but is much greater than across the East African Rift. The tectono-magmatic situation is more akin to Iceland (Section 2) and much of the crust beneath Afar is basaltic through to the Moho. Thus, the whole crustal section has been created since 30 Ma and the area covered is roughly equivalent to the overlap recognized earlier.

**Question 4.11** Alkaline basalts result from low degrees of melting at depth whereas tholeiites tend to be derived from shallower depths by larger melt fractions. Thus, the melting regime has become shallower with time. This can be brought about by extension, reducing the thickness of the lithosphere allowing the underlying mantle to melt by decompression. The orientation of the dykes starts N–S but ends up coast-parallel, implying that the extension that produced the melt is directly related to the extension that produced the Red Sea.

**Question 4.12** For normal continental lithosphere, the relationships shown on Figure 3.27 suggest that the lithosphere needs to be thinned by a factor of almost 4 before melting begins.

**Question 4.13** Large amplitude (>500 nT from peak to trough) anomalies exist over the axial zone and these extend slightly beyond the topographic boundaries of the axial trough (see Figure 4.15a). It can be inferred from this that whatever causes these anomalies extends for a short distance beneath the evaporite cover. These are referred to as type 1 anomalies. Long wavelength anomalies of smaller amplitude (300 nT) occur over the shallow waters that flank the axial trough. These extend from the axial trough to the coastlines on both sides of the Red Sea (type 2 anomalies). Irregular short wavelength, small amplitude (<50 nT) anomalies occur over the continental parts of the section (type 3 anomalies).

**Question 4.14** The large-amplitude anomalies over the axial trough fit remarkably well with the synthetic magnetic anomalies as far back in time as about 5 Ma, but then the correlation breaks down and there is no obvious similarity between the observed and the theoretical anomalies. Note that, as the peaks and troughs of the large amplitude anomalies match, the spreading rate of 1 cm yr$^{-1}$ per flank must be about right. Using a spreading rate of 1 cm yr$^{-1}$ per flank, this means that some 50 km of new oceanic crust has been formed per flank in the past 5 Ma. (Useful note: 1 cm per year = 10 km per Ma.)

**Question 4.15** (a) The crust is thickest under the high areas and thinnest beneath the low areas. The change in thickness is particularly rapid across the escarpments. It should be noted that the topography reflects the changes in crustal thickness in a very subdued way; a change of 10 km in crustal thickness roughly corresponds to a change of 1 km elevation at the surface. (b) The crust under the Afar is attenuated (less than 35 km thick). Only in the south-west corner is this not so. (c) If we take 15 km as the maximum depth to the Moho in an oceanic area, then there is a limited area in the north-central Afar, inside the 16 km contour, that is a possible oceanic crust contender.

**Question 4.16** A–B is a distance of about 79 km. If the pre-rotational widths of each of the four blocks are added up, a total of 12 + 12 + 8 + 37 = 69 km is obtained. So, the total extension is 10 km. The $\beta$ factor is 79/69 = 1.14.

**Question 4.17** The Gulf of Suez is north of the southern extension of the Dead Sea transform fault (Figure 4.3), a conservative plate boundary between the Arabian and the Eurasian Plate. Little if any extension is expected across such a boundary which is characterized by strike–slip movement.

**Question 4.18** (a) The laterite is younger than the underlying Mesozoic sediments but older than the flood basalts. The basalts are dated at 29–30 Ma and the Mesozoic Period ended at 65 Ma. These two dates bracket the age of the laterite. (b) The faults cut through all of the rocks present in the section and so faulting must post-date volcanism (i.e. younger than 29–30 Ma). (c) They are normal faults and so indicate extension.

**Question 4.19**    Rifting was more or less synchronous in the Gulf of Aden and the Red Sea. Therefore, Arabia and Africa acted as rigid plates (see Figure 4.23a).

**Question 4.20**    After closing the Gulf of Suez, there is about 100 km between Africa and Arabia, which means about 100 km of movement of Arabia (south of the Gulf of Aqaba) is needed to close the Red Sea.

**Question 4.21**    The agreement is good. Depending on where you selected the points, movement has been to the north-east between 040° and 050°.

**Question 4.22**    Futurology is an inexact science but the Arabian Plate does not have much more space in which to move. There is little if any oceanic lithosphere remaining on its leading edge and subduction beneath Iran and Iraq has all but ended. This effectively removes the main plate-driving force from the system. On its present track, Arabia will eventually fuse with the slow-moving Eurasian Plate to the north to form one large continental plate, thereby allowing the Red Sea to open a little further. Beyond this, it would require a major re-organization of plate motions for the Red Sea to open wider than 500 km and form an ocean the width of the Atlantic.

**Question 5.1** (a) $2\,000\,000\ km^3$ of basalt were erupted in 1 Ma so the eruption rate is: $2\,000\,000 / 1\,000\,000\ km^3\ yr^{-1} = 2\ km^3\ yr^{-1}$. Published estimates vary from values as low as $0.5\ km^3\ yr^{-1}$ to $8\ km^3\ yr^{-1}$. (b) The highest eruption rate of any plume-related basaltic volcano today is that for Iceland, $0.24\ km^3\ yr^{-1}$. The Deccan basalts represent eruption rates an order of magnitude greater than anything currently active on Earth.

**Question 5.2** First, it is assumed that eruption rates of basaltic magma relate to melt production rates in the mantle. Melt production rates are increased if the lithosphere is thin and the mantle potential temperature is high. Lithospheric extension, even sea-floor spreading, over a mantle plume favours high melting rates and hence high eruption rates.

**Question 5.3** (a) The Réunion basalt has higher abundances of the more incompatible elements (La, Nb and Ba) than the Deccan basalt, but lower abundances of Y and the heavy rare earth elements, Yb and Lu). (b) Higher incompatible elements imply lower melt fractions for the Réunion basalt. Low Y, Yb and Lu indicate residual garnet in the source region.

**Question 5.4** (a) From Figure 2.17, $SiO_2$ between 47% and 49% implies a pressure of melt generation between 1.5–2.5 GPa, equivalent to a depth of 45–75 km. (b) The Moho is located at 35 km beneath most continental crust and so these depths correspond to the mantle. (c) The mantle at these depths is mainly spinel–lherzolite. (d) Continental lithosphere is generally >100 km thick so melt generation at depths of 45–75 km implies an origin within the mantle lithosphere.

# Acknowledgements

Every effort has been made to trace all copyright owners, but if any has been inadvertently overlooked, we will be pleased to make the necessary arrangements at the first opportunity. Grateful acknowledgement is made to the following for permission to reproduce material in this Block:

*Figure 2.2* M. J. Le Bas *et al.* (1986) 'A chemical classification of volcanic rocks based on the total alkali-silica diagram', *Journal of Petrology*, **27**, No. 3, Oxford University Press; *Figure 2.3* G. A. Macdonald and T. Katsura (1964) 'Chemical composition of Hawaiian lavas', *Journal of Petrology*, **5**, No. 1, Oxford University Press; *Figure 2.5* E. Ehlers (1938) 'The interpretation of geological phase diagrams', Yoder and Tilley (eds), W. H. Freeman & Co; *Figure 2.6* E. Ehlers (1962) *Journal of Petrology*, **3**, No. 3, Oxford University Press; *Figure 2.10* E. Takahashi 'Melting of mantle peridotite to 14 GPa', *Journal of Geophysical Research*, **91** © American Geophysical Union; *Figures 3.1, 3.2, 4.12* from US National Intelligence Mapping Agency GTOPO30 digital elevation data, map designed by Steve Drury for the Course Team; *Figure 3.7* B. H. Baker (1987) 'Outline of the petrology of the Kenya rift alkaline province', in J. G. Fitton and B. G. J. Upton (eds), *Alkaline Igneous Rocks*, Geological Society Special Publication; *Figure 3.8* M. Smith. and P. Mosley (1993) 'Crustal heterogeneity and basement influence on the development of the Kenya rift, East Africa', *Tectonics*, **12**, No. 2, American Geophysical Union; *Figure 3.9* J. D. Fairhead (1980) 'The intraplate volcanic centres of North Africa and their possible relation to the East African rift system', *Geodynamic Evolution of the Afro-Arabian Rift System*, **47**, Accademia Nazionale dei Lincei; *Figure 3.10* C. J. Swain (1994) 'Geophysical experiments and models of the Kenya rift before 1989', *Tectonophysics, Special Issue*, Elsevier Science; *Figure 3.12* J. Mechie *et al.* (1994) 'Crustal structure beneath the Kenya rift from axial profile data', Prodehl *et al.* (eds), *Tectonophysics*, **236**, Nos. 1–4, Elsevier Science; *Figure 3.13* Maguire *et al.* (1994) 'A crustal and uppermost mantle cross-sectional model of the Kenya rift, derived from seismic and gravity data', Prodehl *et al.* (eds), **236**, Nos. 1–4, *Tectonophysics*, Elsevier Science; *Figures 3.14, 3.18* W. Verney Green *et al.* (1991) 'A three-dimensional seismic image of the crust and upper mantle beneath the Kenya rift', *Nature*, **354**, Macmillan; *Figure 3.15* KRISP Working Party (1991) 'Large-scale variation in lithospheric structure along and across the Kenya rift', *Nature*, **354**, Macmillan; *Figures 3.16, 3.17* V. Achaver and the KRISP Working Party (1994) 'New ideas on the Kenya rift based on the inversion of the combined dataset of the 1985 and the 1989/90 seismic tomography experiments', Prodehl *et al.* (eds), *Tectonophysics*, **236**, Elsevier Science; *Figure 3.19* Keller *et al.* (1994) 'The East African rift system in the light of KRISP 90', Prodehl *et al.* (eds), *Tectonophysics*, **236**, Nos. 1–4, Elsevier Science;

*Figures 3.21, 3.25* D. B. Hendrie *et al.* (1994) 'Cenozoic extension in Northern Kenya: a quantitative model of rift basin development in the Turkana region', Prodehl *et al.* (eds), *Tectonophysics*, **236**, Elsevier Science; *Figure 3.22* C. J. Ebinger (1989) 'Geometric and kinematic development of border faults and accommodation zones, Kivu–Rusizi rift, Africa', *Tectonics*, **8**, No. 1, American Geophysical Union; *Figure 3.23* M. Smith (1994) 'Stratigraphic and structural constraints on mechanisms of active rifting in the Gregory rift, Kenya', *Tectonophysics*, **236**, Elsevier Science; *Figure 3.24* B. Wernicke (1985) 'Uniform-sense normal simple shear of the continental lithosphere', *Canadian Journal of Earth Sciences*, **22**, National Research Council, Canada; *Figure 3.31* M. Sandiford and D. Coblentz (1994) 'Plate-scale potential-energy distributions and the fragmentation of ageing plates', *Earth and Planetary Science Letters*, **126**, Elsevier Science; *Figures 3.32, 3.33* D. D. Coblentz and M. Sandiford (1994) 'Tectonic stresses in the African plate: constraints on the ambient lithospheric stress state', *Geology*, **22**, Geological Society of America; *Figures 4.7, 4.8* P. Rochette *et al.* (1998) 'Magnetostratigraphy and timing of the Oligocene Ethiopian traps', *Earth and Planetary Science Letters*, **164**, Elsevier Science; *Figures 4.9, 4.10* R. Pik *et al.* (1998) 'The north western Ethiopian plateau flood basalts: classification and spatial distribution of magma types', *Journal of Volcanology and Geothermal Research*, **81**, Elsevier Science; *Figure 4.15* © P. Styles University of Newcastle-upon-Tyne; *Figure 4.16* J. Makris *et al.* (1975) 'Gravity field and crustal structure of North Ethiopia', A. Pilger and A. Rosler (eds), *Afar Depression of Ethiopia*, **14**, Schweizerbart Publishers; *Figure 4.18* S. A. Drury *et al.* (1994) 'Structures related to Red Sea evolution in Northern Eritrea', *Tectonics*, **13**, American Geophysical Union; *Figures 4.20, 4.21* M. Menzies *et al.* (1997) 'Volcanic and non-volcanic rifted margins of the Red Sea and Gulf of Aden', *Geochimica and Cosmochimica Acta*, **61**, Elsevier Science; *Figures 4.23, 4.24* G. I. Omar and M. S. Steckler (1995) 'Fission track evidence on the initial rifting of the Red Sea', *Science*, **270**, AAAS; *Figure 4.25* D. Mackenzie *et al.* (1970) 'Plate tectonics of the Red Sea and East Africa', *Nature*, **226**, Macmillan; *Figure 4.26* S. Joffe and Z. Garfunkel (1987) 'Plate kinematics of the circum Red Sea – a re-evaluation', *Tectonophysics*, **141**, Elsevier Science; *Figure 5.1* R. A. Duncan and M. A. Richards (1991) 'Hotspots, mantle plumes, flood basalts and true polar wander', *Review of Geophysics*, American Geophysical Union; *Figure 5.2* R. W. Griffiths and I. H. Campbell (1990) 'Stirring and structure in mantle starting plumes', *Earth and Planetary Science Letters*, **99**, Elsevier Science.

# Index

*Note:* bold page numbers denote where Glossary terms are introduced/defined